PROGRESS IN CLINICAL AND BIOLOGICAL RESEARCH

Series Editors

Nathan Back   Vincent P. Eijsvoogel   Kurt Hirschhorn   Sidney Udenfriend
George J. Brewer   Robert Grover   Seymour S. Kety   Jonathan W. Uhr

RECENT TITLES

Vol 163: **Prevention of Physical and Mental Congenital Defects**, Maurice Marois, *Editor.* Published in 3 volumes: Part A: *The Scope of the Problem.* Part B: *Epidemiology, Early Detection and Therapy, and Environmental Factors.* Part C: *Basic and Medical Science, Education, and Future Strategies*

Vol 164: **Information and Energy Transduction in Biological Membranes**, C. Liana Bolis, Ernst J.M. Helmreich, Hermann Passow, *Editors*

Vol 165: **The Red Cell: Sixth Ann Arbor Conference**, George J. Brewer, *Editor*

Vol 166: **Human Alkaline Phosphatases,** Torgny Stigbrand, William H. Fishman, *Editors*

Vol 167: **Ethopharmacological Aggression Research**, Klaus A. Miczek, Menno R. Kruk, Berend Olivier, *Editors*

Vol 168: **Epithelial Calcium and Phosphate Transport: Molecular and Cellular Aspects,** Felix Bronner, Meinrad Peterlik, *Editors*

Vol 169: **Biological Perspectives on Aggression,** Kevin J. Flannelly, Robert J. Blanchard, D. Caroline Blanchard, *Editors*

Vol 170: **Porphyrin Localization and Treatment of Tumors,** Daniel R. Doiron, Charles J. Gomer, *Editors*

Vol 171: **Developmental Mechanisms: Normal and Abnormal,** James W. Lash, Lauri Saxén, *Editors*

Vol 172: **Molecular Basis of Cancer,** Robert Rein, *Editor.* Published in 2 volumes: Part A: *Macromolecular Structure, Carcinogens, and Oncogenes.* Part B: *Macromolecular Recognition, Chemotherapy, and Immunology*

Vol 173: **Pepsinogens in Man: Clinical and Genetic Advances,** Johanna Kreuning, I. Michael Samloff, Jerome I. Rotter, Aldur Eriksson, *Editors*

Vol 174: **Enzymology of Carbonyl Metabolism 2: Aldehyde Dehydrogenase, Aldo-Keto Reductase, and Alcohol Dehydrogenase,** T. Geoffrey Flynn, Henry Weiner, *Editors*

Vol 175: **Advances in Neuroblastoma Research,** Audrey E. Evans, Giulio J. D'Angio, Robert C. Seeger, *Editors*

Vol 176: **Contemporary Sensory Neurobiology,** Manning J. Correia, Adrian A. Perachio, *Editors*

Vol 177: **Medical Genetics: Past, Present, Future,** Kåre Berg, *Editor*

Vol 178: **Bluetongue and Related Orbiviruses,** T. Lynwood Barber, Michael M. Jochim, *Editors*

Vol 179: **Taurine: Biological Actions and Clinical Perspectives,** Simo S. Oja, Liisa Ahtee, Pirjo Kontro, Matti K. Paasonen, *Editors*

Vol 180: **Intracellular Protein Catabolism,** Edward A. Khairallah, Judith S. Bond, John W.C. Bird, *Editors*

Vol 181: **Germfree Research: Microflora Control and Its Application to the Biomedical Sciences,** Bernard S. Wostmann, *Editor,* Julian R. Pleasants, Morris Pollard, Bernard A. Teah, Morris Wagner, *Co-Editors*

Vol 182: **Infection, Immunity, and Blood Transfusion,** Roger Y. Dodd and Lewellys F. Barker, *Editors*

Vol 183: **Aldehyde Adducts in Alcoholism,** Michael A. Collins, *Editor*

Vol 184: **Hematopoietic Stem Cell Physiology,** Eugene P. Cronkite, Nicholas Dainiak, Ronald P. McCaffrey, Jiri Palek, Peter J. Quesenberry, *Editors*

Vol 185: **EORTC Genitourinary Group Monograph 2,** Fritz H. Schroeder, Brian Richards, *Editors.* Published in two volumes: Part A: *Therapeutic Principles in Metastatic Prostatic Cancer.* Part B: *Superficial Bladder Tumors*

Vol 186: **Carcinoma of the Large Bowel and Its Precursors,** John R.F. Ingall, Anthony J. Mastromarino, *Editors*

Vol 187: **Normal and Abnormal Bone Growth: Basic and Clinical Research,** Andrew D. Dixon, Bernard G. Sarnat, *Editors*

Vol 188: **Detection and Treatment of Lipid and Lipoprotein Disorders of Childhood,** Kurt Widhalm, Herbert K. Naito, *Editors*

Vol 189: **Bacterial Endotoxins: Structure, Biomedical Significance, and Detection With the Limulus Amebocyte Lysate Test,** Jan W. ten Cate, Harry R. Büller, Augueste Sturk, Jack Levin, *Editors*

Vol 190: **The Interphotoreceptor Matrix in Health and Disease,** C. David Bridges, *Editor,* Alice J. Adler, *Associate Editor*

Vol 191: **Experimental Approaches for the Study of Hemoglobin Switching,** George Stamatoyannopoulos, Arthur W. Nienhuis, *Editors*

Vol 192: **Endocoids,** Harbans Lal, Frank LaBella, John Lane, *Editors*

Vol 193: **Fetal Liver Transplantation,** Robert Peter Gale, Jean-Louis Touraine, Guido Lucarelli, *Editors*

Vol 194: **Diseases of Complex Etiology in Small Populations: Ethnic Differences and Research Approaches,** Ranajit Chakraborty, Emöke J.E. Szathmary, *Editors*

Vol 195: **Cellular and Molecular Aspects of Aging: The Red Cell as a Model,** John W. Eaton, Diane K. Konzen, James G. White, *Editors*

Vol 196: **Advances in Microscopy,** Ronald R. Cowden, Fredrick W. Harrison, *Editors*

Vol 197: **Cooperative Approaches to Research and Development of Orphan Drugs,** Melvin Van Woert, Eunyong Chung, *Editors*

Vol 198: **Biochemistry and Biology of DNA Methylation,** Giulio L. Cantoni, Aharon Razin, *Editors*

Vol 199: **Leukotrienes in Cardiovascular and Pulmonary Function,** Allan M. Lefer, Marlys H. Gee, *Editors*

Vol 200: **Endocrine Genetics and Genetics of Growth,** Costas J. Papadatos, Christos S. Bartsocas, *Editors*

Vol 201: **Primary Chemotherapy in Cancer Medicine,** D.J. Theo Wagener, Geert H. Blijham, Jan B.E. Smeets, Jacques A. Wils, *Editors*

Vol 202: **The 2-5A System: Molecular and Clinical Aspects of the Interferon-Regulated Pathway,** Bryan R.G. Williams, Robert Silverman, *Editors*

Please contact the publisher for information about previous titles in this series.

ADVANCES IN MICROSCOPY

ADVANCES IN MICROSCOPY

Proceedings of the American Microscopical Society Symposium
Held in Philadelphia, Pennsylvania
December 28, 1983

Editors

Ronald R. Cowden
Department of Biophysics
Quillen-Dishner College of Medicine
East Tennessee State University
Johnson City, Tennessee

Frederick W. Harrison
Department of Biology
Western Carolina University
Cullowhee, North Carolina

ALAN R. LISS, INC. • NEW YORK

Address all Inquiries to the Publisher
Alan R. Liss, Inc., 41 East 11th Street, New York, NY 10003

Copyright © 1985 Alan R. Liss, Inc.

Printed in the United States of America

Under the conditions stated below the owner of copyright for this book hereby grants permission to users to make photocopy reproductions of any part or all of its contents for personal or internal organizational use, or for personal or internal use of specific clients. This consent is given on the condition that the copier pay the stated per-copy fee through the Copyright Clearance Center, Incorporated, 27 Congress Street, Salem, MA 01970, as listed in the most current issue of "Permissions to Photocopy" (Publisher's Fee List, distributed by CCC, Inc.), for copying beyond that permitted by sections 107 or 108 of the US Copyright Law. This consent does not extend to other kinds of copying, such as copying for general distribution, for advertising or promotional purposes, for creating new collective works, or for resale.

Library of Congress Cataloging in Publication Data

American Microscopical Society. Symposium (1983 :
  Philadelphia, Pa.)
  Advances in microscopy.

  Includes bibliographies and index.
  1. Microscope and microscopy—Congresses. I. Cowden,
Ronald R. II. Harrison, Frederick William, 1938–
III. Title.
QH201.A33   1983   578'.4   85-13076
ISBN 0-8451-5046-4

Dedication

The editors take pleasure in dedicating this symposium volume to three men: our principal mentors as graduate students, the late Professor Dr. rer. nat., Dr. med. Felix Mainx (R.R.C), the late Professor James T. Penney (F.W.H.), and Professor Martin J. Ulmer under whose presidency of the American Microscopical Society this project was firmly launched.

Professor Mainx, who received his university degrees from the German University in Prague, was deflected from his first interest in Egyptology into the more practical area of plant physiology for his first doctorate. He became involved in *Drosophila* genetics as a result of his fellowship stay at the Kaiser-Wilhelm-Institute in Berlin in the early 30's. By the mid-30's, *Drosophila* genetics was part of the program in the German University in Prague. With the rise of the Third Reich, discretion advised a doctorate in medicine in 1942. After the revolution of 1947, he and his wife arrived as refugees in Vienna. By 1949, an Institute für Allgemeine Biologie was developed for him in the Medical Faculty. In May 1954, one of us (R.R.C.) appeared on his doorstep, very green and somewhat bewildered. He guided R.R.C. through the complexities of the biology of inheritance, and introduced the American to a world of art and history that he would have never otherwise known. Professor Mainx died in July 1983, holding the highest honors that the Austrian government, the University of Vienna, and the Roman Catholic church could confer on a layman.

Professor James T. Penney became a student of the legendary Professor H.V. Wilson at the University of North Carolina at Chapel Hill where Professor Penney established a life-long fascination and involvement with sponges, particularly freshwater sponges. Professor Penney took his first position in the Biology Department of the University of South Carolina at Columbia, and remained there throughout his career. He became immersed in all aspects of freshwater sponge biology: taxonomy, paleontology, ecology, physiology, development and cytology. He also played a large role in the development of the University's athletic program, and at the end of his life, he was, and had been for some years, departmental chairman. His dedication, energy, personality, and example led one of us (F.W.H.) to follow his inspired interest in the biology of freshwater sponges. Professor Penney would probably take some comfort in the fact that many of his "what if" projects have been completed by one of his students. However, perhaps he would be a bit irritated by the fact that other projects in which he placed great stock have not yet received attention.

We owe a great deal to these inspiring professors of another generation; we were proud to have started our doctoral work under their guidance and we salute their memory.

We recognized Professor Martin J. Ulmer, still very much with us, as the President of the American Microscopical Society under whom this symposium project received firm support and sponsorship. He recently retired from the Department of Zoology of Iowa State University in Ames, Iowa. We wish him well and again offer him our thanks.

Ronald R. Cowden, PhD
Frederick W. Harrison, PhD

Contents

Dedication
Ronald R. Cowden and Frederick W. Harrison   vii
Contributors   xi
Introduction
Ronald R. Cowden   xiii
Special Acknowledgments
Ronald R. Cowden and Frederick W. Harrison   xvii

LIGHT MICROSCOPIC METHODS—GENERAL

New Directions and Refinements in Video-Enhanced Microscopy Applied to Problems in Cell Motility
Robert Day Allen.   3

Coordinated Optical and Electron Microscopic Analysis of Particle-Cytoskeletal Interactions in Cytoplasmic Transport
John H. Hayden and Robert D. Allen 13

Computer Assisted Microscope Interferometry by Image Analysis of Living Cells
Jürgen Bereiter-Hahn. 27

Applications of Microspectrofluorometry to Metabolic Control, Cell Physiology and Pathology
Elli Kohen, Joseph G. Hirschberg, and Alexander Rabinovitch 45

Studies on the Role of Ca^{++} in Cell Division With the Use of Fluorescent Probes and Quantitative Video Intensification Microscopy
Jesse E. Sisken, Robert B. Silver, George H. Barrows, and Sally D. Grasch . . . 73

Image Analysis in Biomedical Research
Ronald R. Cowden 89

Scanning Optical Microscopy
Tony Wilson 103

A New Instrument Combining Simultaneous Light and Scanning Electron Microscopy
Cornelia H. Wouters and Johan S. Ploem 115

LIGHT MICROSCOPIC AND ULTRASTRUCTURAL APPROACHES TO THE INVESTIGATION OF CELL NUCLEI AND CHROMATIN

DNA "Standards" and the Range of Accurate DNA Estimates by Feulgen Absorption Microspectrophotometry
Ellen M. Rasch. 137

Refinements in Absorption-Cytometric Measurements of Cellular DNA Content
David C. Allison 167

Microphotometric Demonstration of Differences in Nuclear Structure or Composition
Ronald R. Cowden and Sherill K. Curtis 187

Flow Systems:
    An Editorial Introduction
    Ronald R. Cowden 211

    The AMAC IIIS Transducer
    Robert C. Leif 213

Electron Probe X-Ray Microanalysis Studies on the Ionic Environment of Nuclei and the Maintenance of Chromatin Structure
Ivan L. Cameron 223

ULTRASTRUCTURAL METHODS

Ultrastructural Histochemistry With the High-Voltage Electron Microscope (HVEM)
Eugene L. Vigil, William P. Wergin, and M.N. Christiansen 243

Cathodoluminescence Studies: From Chromatin and Chromosome to Human Hair
Samarendra Basu 265

Ultrasoft X-Ray Historadiography and New Developments in Ultrasoft X-Ray Imaging
    A. Historical Perspective
    Ronald R. Cowden 289

    B. New Approaches to Ultrasoft X-Ray Imaging
    H.M. Epstein 299

Segmentation of Wet Elongated Mitochondria Into Spheres
Samarendra Basu 313

Reflections and Conclusions
Ronald R. Cowden 327

Index 331

Contributors

Robert Day Allen, Department of Biological Sciences, Dartmouth College, Hanover, NH 03755 **[3,13]**

David C. Allison, Department of Surgery, Albuquerque VAMC and the University of New Mexico, Albuquerque, NM 87108 **[167]**

George H. Barrows, Department of Pathology, College of Medicine, University of Louisville, Louisville, KY 40232 **[73]**

Samarendra Basu, Scanning Electron Microscopy Laboratory, New York State Police Crime Laboratory, State Campus, Building 22, Albany, NY 12226 **[265,313]**

Jürgen Bereiter-Hahn, FB Biologie der Johann Wolfgang Goethe Universität, Abteilung: Kinematische Zellforschung, D-600 Frankfurt A.M., Federal Republic of Germany **[27]**

Ivan L. Cameron, Cellular and Structural Biology Department, The University of Texas Health Science Center at San Antonio, San Antonio, TX 78284 **[223]**

M.N. Christiansen, Plant Stress Laboratory, Plant Physiology Hormone Institute, Beltsville Agricultural Research Center, Beltsville, MD 20705 **[243]**

Ronald R. Cowden, Department of Biophysics, Quillen-Dishner College of Medicine, East Tennessee State University, Johnson City, TN 37614 **[vii,xiii,xvii,89,187,211,289,327]**

Sherill K. Curtis, Department of Biophysics, Quillen-Dishner College of Medicine, East Tennessee State University, Johnson City, TN 37614 **[187]**

H.M. Epstein, Battelle Memorial Institute, Columbus, OH 43201 **[299]**

Sally D. Grasch, Department of Pathology, College of Medicine, University of Kentucky, Lexington, KY 40536; present address: Department of Medical Microbiology and Immunology, College of Medicine, University of Kentucky, Lexington, KY 40536 **[73]**

Frederick W. Harrison, Department of Biology, Western Carolina University, Cullowhee, NC 27823 **[vii,xvii]**

John H. Hayden, Department of Biological Sciences, Dartmouth College, Hanover, NH 03755; present address: Department of Biology, Siena College, Loudonville, NY 12211 **[13]**

Joseph G. Hirschberg, Departments of Biology and Physics, Laboratory for Applied Optics and Astrophysics, University of Miami, Coral Gables, FL 33124 **[45]**

Elli Kohen, Papanicolaou Cancer Research Institute, Miami, FL 33101; present address: Department of Biology, University of Miami, Coral Gables, FL 33124 **[45]**

Robert C. Leif, Applied Research Department, Coulter Electronics, Inc., Hialeah, FL 33010 **[213]**

The number in brackets is the opening page number of the contributor's article.

Johan S. Ploem, Department of Histochemistry and Cytochemistry, University of Leiden, 2333 AL Leiden, The Netherlands [115]

Alexander Rabinovitch, Division of Diabetes Research, School of Medicine, University of Miami, Miami, FL 33101 [45]

Ellen M. Rasch, Department of Biophysics, Quillen-Dishner College of Medicine, East Tennessee State University, Johnson City, TN 37614 [137]

Robert B. Silver, Department of Biochemistry, Chicago Medical School, North Chicago, IL 60064; present address: Laboratory of Molecular Biology, University of Wisconsin, Madison, WI 53706 [73]

Jesse E. Sisken, Department of Pathology, College of Medicine, University of Kentucky, Lexington, KY 40536; present address: Department of Medical Microbiology and Immunology, College of Medicine, University of Kentucky, Lexington, KY 40536 [73]

Eugene L. Vigil, Department of Horticulture, University of Maryland, College Park, MD 20742 [243]

William P. Wergin, Plant Stress Laboratory, Plant Physiology Hormone Institute, Beltsville Agricultural Research Center, Beltsville, MD 20705 [243]

Tony Wilson, Department of Engineering Sciences, University of Oxford, Parks Road, Oxford, OX1 3PJ, United Kingdom [103]

Cornelia H. Wouters, Department of Histochemistry and Cytochemistry, University of Leiden, 2333 AL Leiden, The Netherlands [115]

Introduction

Inescapably, microscopes in their various permutations have provided at least an indispensable part of the technology upon which the great three-decade explosion of cell and developmental biology has depended. Just as my father's generation lived to see the Wright Brothers at Kittyhawk and the apex of the Apollo Program with a man walking on the moon, our generation (those of us in our late 40's and 50's) has lived through an age of biomedical research without electron microscopes—or at least in which they were vaguely known as expensive and impractical new experimental designs. It was also a time when *the* research light microscope was an instrument with a light-source placed separately from the microscope, which might have one or two fluorite objectives or exceptionally a set of apochromatic objectives on it, and which was equipped with a large bellows camera or a "camera lucida" for drawing structures visualized by the microscope. On the other hand, one cannot help but reflect that the apochromatic objective has been around since before 1900, and that exotic objectives like the von Rohr monochromatic quartz refracting objectives appeared in the first decade of this century. In short, the resolving power and capacities for correcting the chromatic and spherical aberrations in microscope optics have been available from the start of the century and other special capacities had also emerged. Why then, until the aftermath of World War II, did microscopy seem to crawl into a cave and virtually sleep technologically for five decades? There were experiments to be sure, and this and that system were put together as prototypes, but most of these ideas were never translated into instruments that went onto a dealer's shelf or found their way into a working laboratory.

On reflection, it was always the desire of the microscope manufacturers to sell microscopes, that in the end produced better and more flexible instruments. It seems certain that some feeling swept through the industry in the late 40's and early 50's that the laboratory that already had a "regular" microscope might be persuaded to buy another one with phase contrast optics, special polarized optics, differential interference optics, or one that could be used in microspectroscopy, if these new modes of microscopy expanded the capacities of these laboratories and opened new experimental options for them. The ways to put these new microscopes to good use were being paved.

In Europe, Caspersson in Stockholm, Brachet and Lison in Belgium, and the United Kingdom group at what is now the Medical Research Council Biophysics Unit at King's College (Maurice Wilkins, Howard Davies, P.M.B. Walker and others), as well as Savile Bradbury (then at Oxford) had firmly launched histochemistry, physical microscopy, and even provided the foundations of quantitative cytochemistry. In the United States this technology first surfaced in the 1947 Cold Spring Harbor Symposium, and for at least a decade and beyond a series of A.W. Pollister's students at Columbia—then subsequently their students—filled the literature with exciting revelations based on the arcane practices of histochemistry, physical microscopy and quantitative cytochemistry.

In yet another direction, L.V. Heilbrunn's (University of Pennsylvania) much scorned theories about the role of calcium in cell function and motility were producing those scientists—and their students—who have come to realize that calcium ion movement and interaction with cytoskeletal elements control motility and many other active processes in cells. Perhaps more important to the future of microscopy, this school produced a group of students who needed to look at living cells, and who used light microscopy in its various physical permutations to accomplish this. Professor Shinya Inoue, who by fastidious observation and experimental design—and by his insight that led to the design of rectified polarized light microscope optics—mounted a serious assault on studies of mitotic spindles and microtubule systems in living cells that later extended to all levels of microscopy. Professor R.D. Allen, his co-workers and students, have made very large contributions to this area and it is probably no accident that he and another commanding figure in cell physiology, Professor Daniel Mazia of the University of California at Berkeley, were students of the late Professor Heilbrunn.

Professor H. Stanley Bennett, while Chairman of the Department of Anatomy of the University of Washington in Seattle, probably assembled the nearest approximation of a biophysics department in existence then that did not carry that formal title, and pushed the limits of the microscopy technology of his time. Finally, Professor Britton Chance, in what was then the Johnson Foundation for Medical Physics, probably had just about everyone who considered himself a "biophysicist" at that time through as a graduate student, post-doctoral fellow, staff member or visiting scientist. All areas were open: x-ray diffraction, electrophysiology, chemistry, etc.; but his own interests were in very high-resolution microspectroscopy in living cells. He and his associates made some of the very important contributions to the demonstration of oxidative enzymes in living cells, probes of electrically active membranes, and fluorescence microspectroscopy. Any short summary omits important events and names. One hopes forgiveness for these lapses.

It would be invidious to pass over the parallel development of ultrastructure and the role played by Professors Keith Porter and George Palade at what was then the Rockefeller Institute, and D.W. Fawcett, first at Cornell Medical School, later at Harvard, now working in "retirement" in Kenya. These men carried electron microscopy from a very impractical endeavor in which a successful picture was a very low probability event, to one in which ultrathin sections could be prepared with some regularity, effective fixatives were developed, plastic embedding media were improved, new microtomes and knives were developed, and the electron microscopes themselves were improved. Only one who has pumped a column for 40 minutes, taken 35 plates going through focus to calculate astigmatic compensation, only to see the specimen curl up on the grid, could appreciate how far matters have progressed since 1960. Incidentally, while all these technical advances went forward, the whole world of ultrastructure as we now know it unfolded, disclosing an unknown biological world of fibers, tubes, granules, membranes and amorphous structures.

The point of all this is that advances in technology have been realized by investigators whose determination to better examine or experiment with their material drove them to improve prototypes representing new technology. Their results and publications, coupled with the loud cries of anguish of the scientific community at large who wanted to follow, pushed inherently conservative industrial enterprises into paths that they might otherwise not have elected to follow. In a large sense, the first two decades after World War II saw the introduction of most of the physical modifications and modes of light microscopy that we know today, the packaging of these instruments into more convenient and flexible units, and the introduction of quantitative technology into turn-key units that could be purchased from a dealer. In both transmission and scanning electron microscopy we principally found a steady improvement in convenience: quick pump-down, beam-wobblers, multiple aperture ribbons, substantially better resolution, better decontamination, better stages, more convenient exposure devices, solid-state electronics, and new modes of operation. Coincident with this, the preparative equipment improved to a point at which an individual using reasonable care and judgement could regularly produce useful ultrathin sections and electron micrographs.

This past decade, leaning on the output of the high-technology developments of the space programs, and with new insights into matters physical, chemical and biological based on developments in molecular biology, we have seen an explosion of optics technology at all levels. X-ray microanalysis has arrived in biomedical applications, and image-processing with computers of T.E.M. and S.E.M. pictures has become a standard procedure in many laboratories. We have new forms of physical ultrastructure: NMR microscopy, photoactivation microscopy in which far-ultraviolet light excites electron emissions, x-ray microscopy will certainly arrive, and the realization of the High-Voltage Electron Microscope has allowed the development of three-dimensional perspective without resort to tedious serial sections. Computer reconstruction has been used to simplify three-dimensional reconstructions of serial sections when these are required, and forms of microscopic tomography have come into use. X-ray diffraction equipment and the evaluation of diffraction patterns have improved enormously; and with the introduction of high-flux neutron sources, neutron-diffraction allows placement of hydrogen atoms in maps of crystal structure. The Scanning-Transmission Electron Microscope has allowed resolution that approaches atomic dimensions. It all seems to be falling together.

In light microscopy, the main advances are based on our enhanced ability to quantify and document alterations in composition, organization, or structure in living cells. The computer and image processing have made it reasonably convenient to deal with complex alterations in densities or spectra over short periods of time. Fluorescence microscopy offers enhanced contrast and sensitivity, particularly when combined with video intensification, so that the fate of a single labeled protein molecule can be followed for a day or more within a living cell. Best of all, most of this equipment can be ordered from a regional dealer so that the scientist can focus on biological problems and spend less time fighting the equipment. We know that infrared and ultraviolet light can produce useful fluorescent images, but convenient equipment for these purposes is not yet available. There are polarization and spectral effects yet to be uncovered and understood, and lasers and photosensitive viewing devices have yet to be perfected and integrated into microscopy. Who knows where the next level of ultrastructure will lead?

It is virtually certain that the decade ahead will bring more transfer of technology between physics and engineering on one hand, and the biomedical sciences on the other. The application of physical methods to biological problems has been targeted by a commission of the National Research Council as one of the priority areas of the future. As a biologist, now identified with biophysics, who has lived through the unfolding of histochemistry, physical light microscopy and ultrastructure, one can only rejoice at the innovations we can anticipate. As an invertebrate cytologist and developmental biologist by both training and inclination, I hope that the zoologists, botanists, and microbiologists, as well as the biomedical scientists, will have the insight and the resources to participate in these developments.

This book should offer some insights into the contemporary technology of microscopy, and perhaps some of these methods will find their way into the reader's laboratories.

Ronald R. Cowden
Johnson City, Tennessee

Special Acknowledgments

This book represents the proceedings of a symposium on "Advances in Microscopy," sponsored by the American Microscopical Society at their annual meeting in Philadelphia, December 1983, augmented by some invited contributions mainly from foreign contributors.

We would like to specifically acknowledge the financial support of the following corporations that made the symposium possible:

American Optical Co., (A.O.-Reichert) Buffalo, NY
E.G. and G. Electro-Optics, Salem, MA
E. Leitz, Inc., Rockleigh, NJ
Nikon, Inc. Instrument Division, Garden City, NY
Olympus Corporation of America, Inc., New Hyde Park, NY
Polaroid Inc., Cambridge, MA
Polysciences, Inc., Warrington, PA
Porter Instrument Service, Norcross, GA
Vickers Instruments, Inc., Malden, MA
Carl Zeiss, Inc., Thornwood, NY

We also thank Ms. Sharon Hyder for retyping a number of manuscripts in camera ready format, and the authors for their cooperation. Finally we thank the editors of Alan R. Liss for their patience and help in seeing this book to completion.

Ronald R. Cowden
Frederick W. Harrison

LIGHT MICROSCOPIC METHODS
GENERAL

NEW DIRECTIONS AND REFINEMENTS IN VIDEO-ENHANCED MICROSCOPY APPLIED TO PROBLEMS IN CELL MOTILITY

Robert Day Allen

Department of Biological Sciences
Dartmouth College
Hanover, N.H. 03755

Only a few years ago, almost all research documentation with optical microscopes was by photomicrography. There seemed to be little reason to attach a television camera to a microscope, because television images could convey far less information than still or motion pictures captured on film. The number of pixels (picture elements) in a television image is a fraction of that in even a 16 mm fine grain photomicrograph. Unless the optical image is magnified and the examination area of the specimen reduced so that the numbers of pixels in the two images are matched, the television camera acts as a filter removing the high spatial frequency information in the fine details of an optical microscope image.

Within a surprisingly short time, economic factors, technical innovations and research trends have all but made photographic film nearly obsolete in some laboratories. The biomedical research community has begun to discover the advantages of electronic imaging and digital image processing. These tools were once the exclusive domain of planetary scientists, who were obliged to use electronic images from space telemetry and then digitally image-process them to search for hidden information. Now biomedical scientists are greatly extending the range of information they are able to gather through the microscope by applying both analog and digital processing techniques in electronic imaging. One of the most exciting frontiers in biomedicine is the exploration of phenomena that can now be seen for the first time in electronic images. (For example, see Allen, Metuzals, Tasaki, Brady and Gilbert,

1982; Travis, Kenealy and Allen,1982; Hayden and Allen, 1983).

   Several factors have contributed to the video revolution in optical microscopy. First, the rising price of silver has led a number of scientists to seek substantial savings through videotape or videocassette recording as an alternative to cinemicrography. Second, efficient analog methods for both enhancement and intensification have greatly extended the range of visible structures and processes that can be studied with the optical microscope (Allen and Allen, 1983 a,b; Reynolds and Taylor, 1980). Third, research trends have created a demand for instruments that can probe living cells to see finer detail, detect smaller structures, and localize a limited number of fluorescent molecules serving as markers for enzymes, cytoskeletal proteins, membrane receptors, or viral genes.

   It is important to point out that there exists an impressive (and confusing) array of commercially available television cameras, components and accessories that vary greatly in price, quality, and compatibility with one another. Almost any camera can be used to record protists swimming around on a slide at low power under a phase contrast microscope. The expenditure need not be more than several hundred dollars. However, specialty cameras for high speed video recording, enhancement cameras, and intensification cameras may cost from 10-50 times more, depending on their purpose, complexity, and performance specifications.

VIDEO INTENSIFICATION MICROSCOPY

   Video intensification is the term that describes the process by which an image, formed by photons that are too few in number for the scene to be visible by eye, is made visible through some analog process that multiplies the photoelectrons available at the photosensitive surface of the camera (see Reynolds and Taylor,1980). In the simplest case, some intensification can be achieved by amplifying the signal from a standard video camera. Most of the sensitive camera designs currently in use for intensification were by-products of military technology from the Vietnam War. They include the silicon intensified target (SIT) camera, single-, double-, and triple-intensified vidicons, the intensified SIT cameras, and cameras containing combinations of 1-3 microchannel plates. Soon to be released in this country will be a photon-counting camera that stores single photons positionally in the display of a frame memory 16 bits deep (Hamamatsu Systems, Inc.).

VIDEO-ENHANCEMENT

Broadly speaking, enhancement of an image means to bring out its hidden features. In general, television cameras enhance images with optical microscopes to some extent merely by virtue of their linearity in comparison with the response of the naked eye, which is logarithmic. The strategy of enhancement is to amplify the video signal to emphasize the AC details riding on top of the video signal and then to depress the DC level of the signal with offset to reduce brightness and therefore increase contrast on the monitor. In the microscope, these operations are best performed when the compensator of a polarizing or differential interference microscope has been adjusted at too high a level for the human eye, in the range of $\lambda/9-\lambda/4$. Although the visual image in the microscope virtually vanishes, the enhanced television image now contains information that could not have been seen by eye nor captured on photomicrographs. This is the principle of the AVEC methods, which are mainly responsible for the enhancement side of the video revolution in microscopy (Allen, Travis, Allen and Yilmaz,1981; Allen, Allen and Travis, 1981; Allen and Allen, 1983a,b).

NEW DEMANDS PLACED ON MICROSCOPE DESIGN

Most of the optical microscopes available from vendors today were designed with visual observation and photomicrography in mind. Because of the diffraction limitation on optical resolution, it was assumed in the design specifications that useful magnifications beyond 2,000X would be unnecessary, and that specimens smaller than 0.2 μm would never be viewed.

The video revolution has invalidated those assumptions, so that many of the "modern" microscopes are to one degree or another obsolete for videomicroscopy. A notable exception is the Zeiss Axiomat, considered by some to have been a "white elephant" when it was introduced, because of its size, complexity and cost. When we consider the list of desirable features for a suitable videomicroscope, it will become apparent how prescient was its designer, Dr. Kurt Michel (1974), for whom this was the dream research microscope for the last quarter of the 20th century. His assessment was correct, but for some reasons he could not have predicted at the time.

The most important requirement for a videomicroscope is mechanical stability. At a magnification of 10,000-20,000 any vibration transmitted to the specimen is highly undesirable. So is any sticking or backlash of the stage controls, or loss of centration of a rotating stage. In general, the mechanical stability of the Axiomat is about ten times better than that of the less massive (and less costly) stands.

The second requirement is excellent image quality at highest magnification. Only a few major optical companies manufacture planapochromatic objectives which give the best images. It is important in television imaging to fill the screen with the features of the specimens that are of special interest. It is desirable for this reason to have a panchratic or "zoom" lens to make this conveniently possible. It is not possible to have a high quality zoom unless the objective is internally corrected so that it does not require compensating oculars of fixed magnification. The planapochromats of the Axiomat are internally corrected, and the observation module contains a zoom lens. Recently Nikon has produced a prototype of a zoom projective ocular which is expected to be available commercially. Although the objectives are not entirely internally corrected, the images obtained with this zoom ocular are quite acceptable. Unfortunately, this zoom projective ocular is probably not suitable for use with objectives of different manufacturers. Many microscope manufacturers have been slow to introduce any form of suitable optical interface for TV cameras. Therefore, buyers should be careful to check design specifications and see what accessories are available before purchasing a microscope for videomicroscopy.

When video-enhancement is pushed to its limits, for

example to observe cytoskeletal elements in living cells, two additional limitations arise. The first and most important is mottle, a pattern of detail that remains in the image when the specimen is moved out of the field or out of focus. Mottle is apparently the sum of all disturbances caused by any dirt of surface imperfections in any of the lenses of the entire optical system of the microscope (Allen and Allen, 1983a). It is usually not visible or detectable photographically, but seen only in the electronic image when extreme video-enhancement is carried out. The second limitation is lack of absolute homogeneity of illumination. In visual microscopy and photomicrography it is usually enough to adjust the microscope for Köhler illumination and to center the light source carefully. At very high degrees of enhancement, the collector and condenser lens corrections and the source structure are often inadequate to provide sufficiently homogeneous illumination. This limitation in general is more serious in the less expensive microscopes, and is an important factor to be considered in the choice of a microscope. It is noteworthy that the Zeiss Axiomat with a well centered 50 watt mercury lamp will illuminate the specimen with acceptable homogeneity even with a gain of 16 times applied to the video signal. The other microscopes we have tried have not performed as well. The mottle subtraction process (vide infra) serves also to even out some of the inhomogeneity in illumination over a limited range.

THE USE OF FRAME MEMORIES

A frame memory (frame store) is an electronic device for holding and manipulating an image. The only useful frame memories for optical microscopy are those that are capable of real-time arithmetic operations. These devices can greatly extend the capability of a videomicroscope (Allen and Allen, 1983a). The most important operations that can be performed include the digital subtraction of mottle and the digital averaging of images to reduce pixel noise.

Mottle subtraction effectively removes the disturbing pattern seen whether the specimen is in focus or not. This operation contributes to the clarity of the image in the same way that cleaning the windshield of a vehicle improves the image seen through it.

Averaging of pixel noise removes the "salt and pepper" appearance of single frame images. One chooses an appropriate time interval over which to average. One hundred frames, for example, would reduce the noise by the square root of 100, or 10. Several companies advertise frame memories: Quantex, Imaging Technology, Colorado Video, Hamamatsu, etc. Some are software operated, others are hard-wired. In general, those that are hard-wired can be put into operation more rapidly. Frame memories are available with different dynamic ranges. In our experience, one that is 8 bits deep is marginal for microscopy, where extended grayscale is as important as resolution.

A PHOTONIC MICROSCOPE SYSTEM

In 1981, we could see that a system could be designed in which the television components were optimized for the kinds of tasks microscopists do. It was also clear that some changes would have to be made in microscopes for the sake of certain functions that the system as a whole could perform.

Early in 1983 I entered into a collaboration with Hamamatsu Photonics, K.K. of Japan and Carl Zeiss/Oberkochen of West Germany to produce the first "Photonic Microscope System", designed from the light source to the monitor screen to provide optimal microscope images. Through the kind cooperation of Mr. Michael Hiller of Carl Zeiss/Oberkochen, we obtained a "cut down" video version AXIOMAT with certain modifications and lacking the photomicrographic module which is unnecessary if images are recorded electronically. High quality photographs - "videomicrographs" - can be captured from the monitor with an ordinary 35 mm SLR camera using a Ronchi-ruling in front of the camera lens to filter out the TV lines (Inoué, 1981). This version of the AXIOMAT is more economical to purchase yet has all the desirable attributes of the more expensive complete model.

The television components of the Photonic Microscope System consist of a TV camera with high dynamic range **(Chalnicon, plumbicon and Saticon are preferred), a** control board that operates a triple frame memory 16 bits deep. The instrument is called the C-1966 AVEC/VIM because its basic design incorporates features required for both Allen video enhanced contrast (AVEC) and video intensification microscopy (VIM).

The control board has camera controls, analog (AVEC) enhancement controls, and controls that permit the operator to subtract mottle digitally to clarify the image, and then to select an appropriate interval over which to average signals to reduce noise. These are the basic functions that the microscopist will use for video-enhancement.

For video intensification, photon noise and pixel noise are distinct problems at low light fluxes. Therefore, summing and averaging video frames are the most frequently desired arithmetic operations.

OTHER OPERATIONS

In principle, the frame memory of the C-1966 can be programmed to carry out almost any arithmetical operation. A few of the more useful ones have been hard-wired for ease and rapidity of operation.

1. Sequential subtraction to detect motion

Images are sequentially subtracted over a variable interval so that only motions or dimensional changes are visible while constant features are removed. This mode was intended for use in the study of fast axonal transport and other motility phenomena.

2. "Trace"

In this mode, the sequentially subtracted images are summed to show the trajectories of moving elements.

3. "Differential"

This mode presents the derivative of the raw video image. In phase contrast or bright-field, this provides a differential (shadow-cast) image like that obtained in differential interference contrast. Many of the undesirable characteristics of the phase contrast image can be removed in this way.

4. "Filtering"

In this mode the image is "smoothed" by arithmetic operations on near-neighbor pixels. This operation may be helpful in smoothing certain fluorescent images so that they become more easily interpretable.

THE FUTURE OF PHOTONIC MICROSCOPY

The C-1966 AVEC/VIM was intended to be a basic system design that would be expanded and updated as the requirements of microscopists become more sophisticated.

For example, in the future it will be possible to observe the same specimen simultaneously with AVEC and VIM using a dual control board and two cameras with the two images in register. With such a device, for example, an experimenter could identify receptor sites or cytoskeletal elements while recording a physiological process in which they took part.

As with all revolutions, microscopy will never be the same. Microscopists must now not only understand the optical aspects of their instruments and methods, but also the mysteries and advantages of electronic imaging.

REFERENCES

Allen RD, Allen NS (1983a). Video-enhanced microscopy with a computer frame memory. J Micr 129:3.
Allen R, Allen N (1983b). Method of adjusting a video microscope system incorporating polarization or interference optics for optimum imaging conditions. U.S. Pat. 4,412,246.
Allen R, Allen N, Travis JL (1981). Video-enhanced contrast differential interference contrast (AVEC-DIC) microscopy: a new method capable of analyzing microtubule-related movement in the reticulopodial network of Allogromia laticollaris. Cell Motil 1:291.

Allen R, Travis JL, Allen NS, Yilmaz H (1981). Video-enhanced contrast polarization (AVEC-POL) microscopy: a new method applied to the detection of birefringence in the reticulopodial network of Allogromia laticollaris. Cell Motil 1:275.

Allen R, Metuzals J, Tasaki I, Brady ST, Gilbert SP (1982). Fast axonal transport in giant squid axon. Science 218:1127.

Hayden JH, Allen R (1983). Cytoplasmic transport in keratocytes: direct visualization of particle translocation along microtubules. Cell Motil 3:1.

Inoué S (1981). Video image processing greatly enhances contrast quality, and speed in polarization-based microscopy. J Micr 89:346.

Michel K (1974). Zeiss Axiomat. A new concept in microscope design. Zeiss Information 21:1.

Reynolds GT, Taylor DL (1980). Image intensification applied to light microscopy. Biosci 30:586.

Travis JL, Kenealy JFT, Allen RD (1983). Studies on the motility of the foraminifera 11. The dynamic microtubular cytoskeleton. J Cell Biol 97:1668.

COORDINATED OPTICAL AND ELECTRON MICROSCOPIC ANALYSIS OF
PARTICLE-CYTOSKELETAL INTERACTIONS IN CYTOPLASMIC TRANSPORT

John H. Hayden and Robert D. Allen

Department of Biological Sciences
Dartmouth College
Hanover, N.H. 03755

Recent advances in video microscopy have made it possible to obtain images of increased contrast and resolution (Allen et al.,1981 a+b; Allen and Allen, 1983). Analog enhancement of contrast is generated by optical and electronic gain. The optical gain is a result of the high compensator setting of $\lambda/9$ which increases the difference between the intensity of the specimen and the background but floods the field with stray light. This stray light can be removed by electronic offset. Contrast is further amplified by electronic gain control in the camera, while variable offset further reduces background brightness. By using a computer frame memory to store background from lens defects, and by continuously subtracting it from the live image, a further gain in image quality is realized as the object appears against a homogeneously illuminated background. These techniques, especially when used in conjunction with immunofluorescence (Hayden et al.,1983) or whole mount electron microscopy (Hayden and Allen, 1983; Hayden and Allen, in preparation), provide a powerful tool for the study of a number of biological problems.

In the work to be described here, these techniques were applied to the problem of cytoplasmic transport of optically detectable organelles and vesicles (collectively called particles). Rebhun (1963, 1972) was the first to recognize the transport of particles as a distinct process, and he gave it the name saltation to distinguish it from random Brownian motion. Microtubules have been indirectly implicated as part of the saltation process. The saltation path and/or the orientation of the particles are consistent with the pattern of organization of the microtubules as seen with electron microscopy (Freed and Lebowitz, 1970; Wagner and Rosenberg, 1973; Wang and Goldman, 1978) or immunofluorescence microscopy (Heggenness et al., 1978; Ball and Singer, 1982; Couchman and Rees, 1982). Microtubule depolymerizing agents inhibit saltation in many cell types (Rebhun, 1972; Hammond and Smith, 1977; Freed and Lebowitz, 1970). However, in these studies it was never possible to demonstrate directly that particle transport was associated with microtubules because the cytoplasmic microtubules could not be detected.

Through the use of Allen video enhanced contrast DIC (AVEC-DIC) microscopy with a computer frame memory, microtubules in a slurry have been observed as distinct linear elements (Allen and Allen, 1983). We decided that by using AVEC-DIC microscopy enough contrast should be generated to detect cytoplasmic microtubules in a suitable cell culture model. Keratocytes from the corneal stroma of Rana pipiens were found to grow as very thin and flattened cells on glass cover slips. These fibroblasts were examined with AVEC-DIC microscopy using a Zeiss Axiomat equipped with a 100x/1.3 NA planapochromatic objective. A bias retardation of $\lambda/9$ was used. The image was projected onto a C-1000-01 chalnicon camera equipped with analog enhancement, and driven by a model 1400 Polyprocessor Frame Memory (Hamamatsu Systems Inc., Waltham, Mass.). The computer frame memory was used to subtract mottle from the live image. In sufficiently thin regions of these cells, a network of linear elements that intersected one another at various points and angles was detected (Fig. 1). All particle transport was observed to take place along these linear elements. Two examples of particle transport along linear elements are shown in Figure 3.

The problem at this point was to determine the protein composition of these linear elements. This was accomplished by performing the indirect immunofluorescence technique, using anti-tubulin IgG, while the cells remained in place and in focus on the microscope. To allow entry of the IgG's, the cells were first lysed in a microtubule stablizing solution containing 0.5% Triton X-100 (Hayden et al., 1983). They were then fixed for 10 minutes in 3.7% formaldehyde, washed in phosphate-buffered saline (PBS), incubated for 1/2 hour in rabbit anti-tubulin IgG, washed in PBS, incubated for 1/2 hour with goat anti-rabbit IgG conjugated to rhodamine, and washed with PBS. The Axiomat was then switched to epifluorescence optics and the cells were viewed with a silicon intensified target (SIT) camera (Model C-1000-12, Hamamatsu Systems Inc., Waltham,

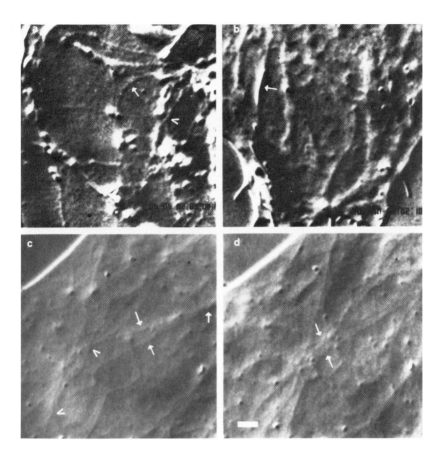

Fig. 1. a-c) AVEC-DIC images of three cells showing the linear elements present. The images of cells a+b were recorded at a video gain of 16, whereas images of the cell in c+d were recorded at a gain of 8, hence the variation in contrast. Note the curved linear element at arrow in (a), the straight linear elements in (c) that converge into one (arrows), and the regions where linear elements intersect (arrowheads). In (b) an elongate particle (arrow) is shown that was in the process of moving down a linear element. d). AVEC-DIC image of same field in (c) after 4 min have elapsed. Note that the linear elements in the center (arrows) appear to have moved closer together. Bar = $2 \mu m$.

Mass.). The averaging mode of the computer frame memory was used to increase the signal to noise ratio.

The cell shown in Figure 1d is shown in Figure 2a after lysis. Note that the linear elements remained in place and gained in contrast upon lysis. The gain in contrast is probably due to the removal of soluble components upon cell lysis, thereby increasing the refractive gradient between the linear elements and the background. In Figure 2 b+c the cell is shown after treatment with rabbit anti-tubulin IgG and goat anti-rabbit IgG conjugated to rhodamine, respectively. Note that the linear elements gained further in contrast as they gained in dry mass concentration. Figure 2d shows that the linear elements identified by anti-tubulin immunofluorescence correspond exactly in number, form, and position to those seen in the live cell. Thus, the linear elements along which particles move are composed of microtubules. Since it is not possible at the level of resolution of the light microscope to determine whether these linear elements are composed of single microtubules or bundles, we call these linear elements microtubular linear elements (MTLE's).

A number of interesting observations were made when the paths and behaviors of the particles were compared to the corresponding images of the MTLE's. A particle, as it moves across the cell, can switch from one MTLE to another one intersecting it. Figure 3 a-c shows a small spherical particle which moved up a linear element and paused. It then turned sharply and moved to the right on another linear element. Figure 3 d-f shows an elongate particle which moved down a linear element and paused. It then turned sharply and moved along the same linear element as the small spherical particle. When the particle paths are compared to the corresponding MTLE's (Fig. 2 c-d) it appears that the particles switched MTLE's during their translocation.

Particles can also move at different rates even on the same MTLE. The small spherical particle moved up the linear element, shown in Figure 3 b+c, and off the screen to the right at 2.1 microns/sec.. The elongate particle, shown in Figure 3 e+f, moved up the same linear element and off the screen to the right at 1.0 microns/sec.. The speed, pause time, and distance moved can vary even for the same particle as it moves along a MTLE.

Particles can move in either direction on a single MTLE.

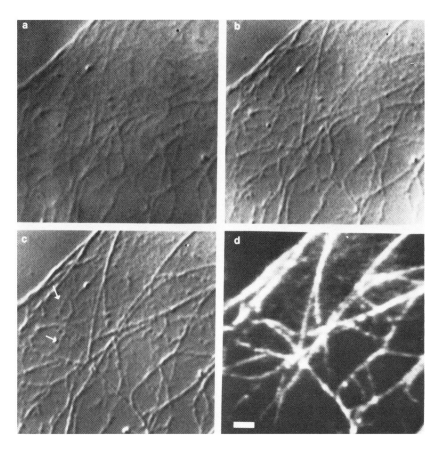

Fig. 2. a) AVEC-DIC image of the same cell shown in Figure 1 after lysis in microtubule stabilizing solution. b) AVEC-DIC image of cell after incubation with rabbit anti-tubulin. c) AVEC-DIC image of cell after incubation with goat anti-rabbit IgG. Note gain in contrast of linear elements after each antibody treatment. d) Indirect immunofluorescent image of cell shown in a-c. Note that the fluorescent linear elements correspond to the linear elements seen in the live cell. Note also that after lysis some less contrasting linear elements (arrows) become apparent. However, these do not stain with anti-tubulin. Bar = 2 µm.

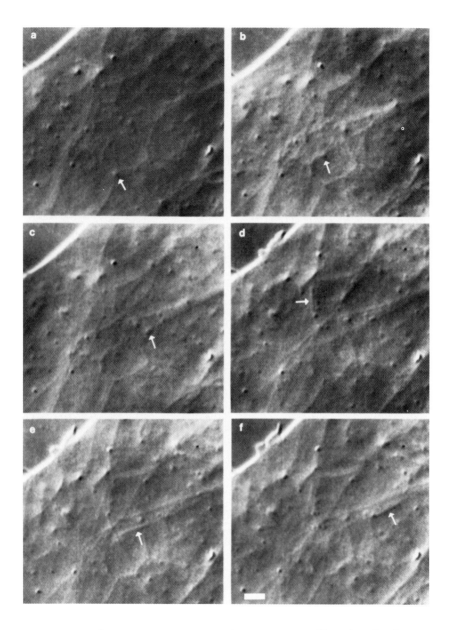

Fig. 3. a-c) AVEC-DIC images showing a small spherical particle (arrow) that moved upward toward the center of the field and then turned right and moved along a straight MTLE.

Fig 3. (continued from previous page). d-f) AVEC-DIC image of an elongate particle (arrow) that moved down a linear element and then turned to follow the same MTLE as the small spherical particle shown in a-c. These movements occurred in the same cell in which the MTLE's are delineated by anti-tubulin immunofluorescence (Fig. 2). When compared to Figure 2 c-d it appears that these particles switched MTLE's when they changed directions. Both particles paused as they switched MTLE's. Bar = 2 μm.

Figure 4 a-c shows a particle moving up a linear element, pausing, and then returning to its original position. When this movement is compared to the corresponding fluorescent image (Fig. 4e) it appears that the particle moved bidirectionally on a single MTLE. Figure 5 a-d shows an example of two particles moving in opposite directions on the same linear element. Figure 5 a-b shows a particle moving up a curvalinear element. Figure 5 c-d shows a second particle moving down the same element about one minute later. When the movements are compared to the corresponding fluorescent image (Fig. 4e) it appears that the two particles moved in opposite directions on the same MTLE.

There appears to be more than one site for the force generator on a given particle. Figure 6 a-d shows an elongate particle moving down a linear element until it comes to an intersection with another linear element. It then forms an inverted Y as portions of the particle move down both linear elements. The particle then forms an inverted U and finally becomes aligned along the second linear element.

Finally, particles are very likely anchored to the cytoskeleton when not transported since they show no Brownian motion.

The observations made using AVEC-DIC in conjunction with immunofluorescence microscopy are summarized below:
1. Using AVEC-DIC microscopy a network of linear elements is seen in corneal keratocytes from Rana pipiens.
2. These linear elements were directly identified as microtubules (MTLE's) by anti-tubulin immunofluorescence.
3. Particle translocation takes place along MTLE's.
4. Particles can switch from one MTLE to another as they move across the cell.
5. One or more particles can move bidirectionally on an MTLE.

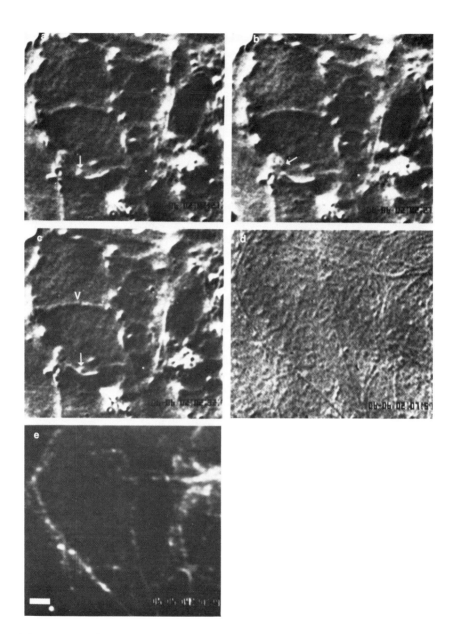

Fig. 4. a-c) AVEC-DIC images of a particle (arrows) that moved along an MTLE (a+b), paused (b), and then reversed direction (c). d) AVEC-DIC image after lysis of cell. e) corresponding

fluorescence micrograph. Note that a linear element in (c) (arrowhead) is not represented in (d) or (e). Bar = 2 µm.

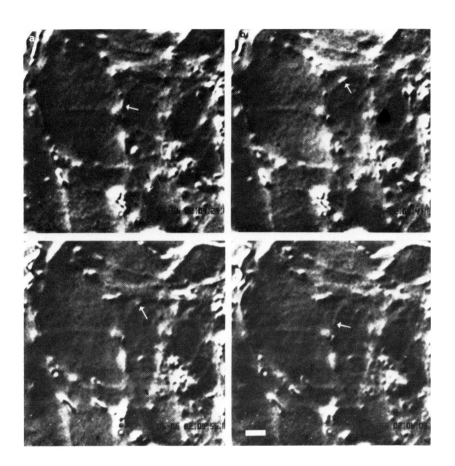

Fig. 5 a-b) images showing curved MTLE along which a particle moved (arrows). c-d) About 1 min later a small particle (arrows) came down the same curved MTLE in the opposite direction. The corresponding fluorescence micrograph is shown in Figure 4e. Bar = 2 µm.

6. Two particles can move at different rates on the same MTLE.
7. The speeds, pause times, and distances moved can vary even for the same particle as it moves across the cell.
8. Particles seem to be attached to the cytoskeleton when

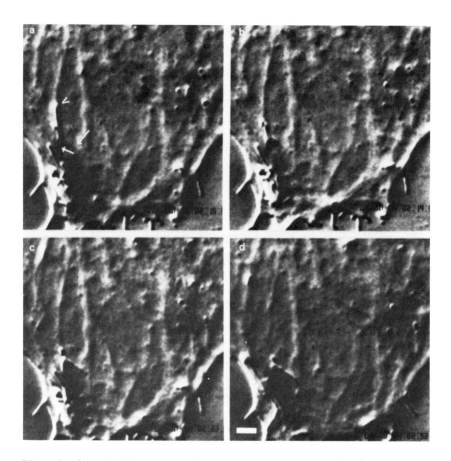

Fig. 6. AVEC-DIC images of an elongate particle (arrowhead) that moved down a linear element. a) The particle moved past the intersection of two linear elements (arrows). b,c) As the particle passed the intersection it was deformed into an inverted "Y" as parts of it followed both linear elements. d) Finally, the entire particle became aligned along the second linear element. Bar = 2 µm.

not moving, since they show no Brownian motion.
9. There appears to be more than one site for the force generator on a given particle.

Our next objective was to identify at an ultrastructural level what we were detecting with AVEC-DIC in the living cell.

Our aims were to determine whether:
1. Single microtubules could be detected in the living cell.
2. Particles could move along single microtubules.
3. Particles could move bidirectionally on single microtubules.

To accomplish this AVEC-DIC microscopy was used in conjunction with whole mount electron microscopy. The keratocytes were grown on Formvar- and carbon-coated gold London Finder grids. The grids were mounted on glass cover slips and then examined on the Zeiss Axiomat with AVEC-DIC microscopy as described earlier. One problem with using electron microscopy grids coated with plastic for light microscopy is that they introduce considerable stray light into the image. This makes it more difficult to detect the MTLE's. However, by using the background subtraction mode of the computer much of the effect of this stray light is removed from the final image.

Once an appropriate cell was found, it was videotaped until fixation was completed in a microtubule stabilizing solution containing 1.0% glutaraldehyde. The preparation was postfixed in 1.0% osmium and then processed for whole mount electron microscopy essentially according to the method of Wolosewick and Porter (1976). The region of the cell recorded on the videotape was located on the electron microscope and photographed.

Preliminary observations demonstrate the following:
1. Single microtubules can be detected in the thin portions of living cells.
2. Single 10 nm filaments can be detected in the lysed cell.
3. Particles can move along single microtubules.

Experiments are in progress to determine whether particles can move bidirectionally on a single microtubule. This is an interesting question when one considers mechanisms of particle transport. If the motive force were delivered by a microtubule-associated mechanochemical transducer such as dynein, then one would predict only unidirectional transport on a single microtubule. On the other hand, if the mechanochemical transducer is bound to the particles, then particles might move bidirectionally with or without changes in orientation. Variations in velocity or direction could be explained by the coming into play of different numbers of transducer molecules with the same or different polarities. Experiments are also planned to correlate a given particle behavior at the time of fixation to its cytoskeletal relationships as seen with stereo electron micrographs. This should lead to new insights into the cytoskeletal elements involved in particle transport.

In summation, AVEC-DIC microscopy with a computer frame memory has allowed us to detect cytoskeletal structures as small as microtubules in the living cell. Hence, we have been able, for the first time, to identify some of their functions. This work is continuing and in future work a Zeiss Axiomat with UV optics will be utilized. This will give us a doubling of resolution which should help immensely in our study of the complex interactions of cytoskeletal components manifested in the various types of cell motility.

LITERATURE CITED

Allen RD, Allen NS (1983). Video-enhanced microscopy with a computer frame memory. J of Microscopy 129:3.

Allen RD, Allen NS, Travis JL (1981a). Video-enhanced contrast, differential interference contrast (AVEC-DIC) microscopy: A new method capable of analyzing microtubule-related motility in the reticulopodial network of Allogromia laticollaris. Cell Motility 1:291.

Allen RD, Travis JL, Allen NS, Yilmaz H (1981b). Video-enhanced contrast polarization (AVEC-POL) microscopy: a new method applied to the detection of birefringence in the motile reticulopodial network of Allogromia laticollaris. Cell Motility 1:275.

Ball EH, Singer SJ (1982). Mitochondria are associated with microtubules and not with intermediate filaments in cultured fibroblasts. Proc Natl Acad Sci USA 79:123.

Couchman JR, Rees DA (1982). Organelle-cytoskeleton relationships in fibroblasts: mitochondria, golgi apparatus, and endoplasmic reticulum in phases of movement and growth. European J of Cell Biol 27:47.

Freed JJ, Lebowitz MM (1970). The association of a class of saltatory movements with microtubules in cultured cells. J Cell Biol. 45:334.

Hammond GR, Smith RS (1977). Inhibition of the rapid movement of optically detectable axonal particles by colchicine and vinblastine. Brain Research 128:227.

Hayden JH, Allen RD (1983). Direct visualization of single microtubules in living cells. J Cell Biol 97:5a.

Hayden JH, Allen RD. Direct visualization of particle motion along single microtubules. In preparation.

Hayden JH, Allen RD, Goldman Rd (1983). Cytoplasmic transport in keratocytes: Direct visualization of particle translocation along microtubules. Cell Motility 3:1.

Heggenness MH, Simon M, Singer SJ (1978). Association of mitochondria with microtubules in cultured cells. Proc Natl Acad Sci USA 75:3863.

Rebhun LI (1963). Saltatory particle movements and mitosis. In Levine L (ed): The Cell in Mitosis, New York: Academic Press, p 67.

Rebhun LI (1972). Polarized intracellular transport: Saltatory movements and cytoplasmic streaming. Int Rev Cytol 32:92.

Wagner RC, Rosenberg MD (1973). Endocytosis in Chang liver cells: The role of microtubules in vacuole orientation and movement. Cytobiologie 7:20.

Wang E, Goldman RD (1978). Functions of cytoplasmic fibers in intracellular movements in BHK-21 cells. J Cell Biol 79:708.

Wolosewick JJ, Porter KR (1976). Stereo high-voltage electron microscopy of whole cells of human diploid line, WI-38. Am J Anat 147:303.

COMPUTER ASSISTED MICROSCOPE INTERFEROMETRY BY IMAGE ANALYSIS OF LIVING CELLS

Jurgen Bereiter-Hahn
FB Biologie der Johann Wolfgang Goethe-
Universität, AK Kinem. Zellforschung
Frankfurt A.M., FRG

INTRODUCTION

Interference microscopy provides information on refractive index and thickness of an object of microscopical dimensions. Because it is a non-destructive method it can be used to follow these parameters in living cells without causing physiological disturbances.

The primary information provided by an interference microscope is a modulation in local light intensity caused by interference of light beams which differ in phase due to different optical paths: One penetrates the object under investigation, while the other serves as a reference beam. This basic principle is independent of the type of microscope used (for details see Ross, 1967 or Beyer, 1973).

The optical path difference Γ depends on specimen thickness (d) and difference between refractive index of the specimen (n_c) and its environment (n_m):

$$\Gamma = (n_c - n_m) \cdot d \qquad (1)$$

Specimens which introduce optical path differences of $\lambda/2$ or integer multiples of $\lambda/2$ appear dark due to destructive mutual extinction of object and reference beam by interference, while those causing λ or integer multiples of λ path difference, appear bright due to constructive interference. The order of interference, and therefore the range of Γ, has to be determined with white light illumination by color discrimination, while exact measurements have to be undertaken using monochromatic light. The refractive index of a cell (n_c) or an organelle is related to its concentration of solid matter (c) (= dry mass concentration) by the specific refraction increment α (Barer & Joseph, 1954). α (cm^3/g) represents the increase in the refractive index of a solution caused by raising solid matter concentration by 1% (1g/100cm^3).

$$c = \frac{n_c - n_w}{100\ \alpha} \quad (2)$$

n_w = refractive index of water.

For substances of biological significance α is between 0.0015 and 0.0019 cm^3/g (Barer & Joseph), 1954. Using equation (1) and (2): solid concentration c is given by:

$$c = \frac{\Gamma}{100\ \alpha \cdot d} \quad (3)$$

Dry mass (DM) of a specimen can be calculated, when the area (A) and the related optical path difference are known:

$$DM = \frac{\Gamma \cdot A}{100\ \alpha} + \frac{(n_m - n_w)\ V}{100\ \alpha} \quad (4)$$

For small differences between the refractive index of the surrounding medium and the refractive index of water, the second part of this equation becomes very small, so it can be neglected. Volume (V) determination is based on the measurement of area (A) and thickness of a specimen according to formula (1). For this purpose the refractive index of the optical object has to be determined by immersion refractometry (Barer & Ross, 1952, Barer & Joseph, 1954, 1955) or by measurement of Γ either at two different wavelengths (e.g. blue and orange or green light) or in suspensions of media of different refractive indices. For practical use with living cells the latter two methods are not recommended because of the hazardous effects of blue light on living cells (Beck & Bereiter-Hahn, 1984), and the disturbances caused by a change of immersion medium.

An example for the determination of each of these three parameters in living cells by using computer-aided microinterferometry based on image analysis will be given, with a short comment on the physiological significance:

1) Dry mass concentration in different regions of a developing sea urchin egg;
2) Dry mass of tissue culture cells, their nuclei and nucleoli;
3) Volume of single epidermal cells under various osmotic conditions.

DETERMINATION OF DRY MASS CONCENTRATION IN A SEA URCHIN ZYGOTE

The first cleavage of sea urchin eggs (<u>Psammechinus miliaris</u> Gmel) has been investigated by means of microinterferometry using a Leitz interference microscope of the Mach-Zehnder type (Bereiter-Hahn, 1971). The sequence of events is documented on 35mm film negatives. Solid concentration is calculated with a computer from densitograms taken with a Joyce-Loebl-micro-densitometer using formula (6). Recent developments in image analysis allow rapid and convenient collection of data. The following procedure is based on the facilities of the interactive image analysis system (IBAS) developed by Kontron (purchased by Zeiss, Oberkochen F.R.G.).

Statement of the problem: Sea urchin (<u>Psammechinus miliaris</u> Gmel) eggs are of spherical shape with a diameter of 0.15-0.2 mm (Fig. 1). The question of scientific interest is whether the solid matter is evenly distributed throughout the cell or whether zones with different dry mass concentration appear, probably related to the cleavage process.

Therefore the sphere is assumed to be virtually composed of a series of concentric shells (Fig. 2). When immersed in sea water, sea urchin eggs produce an optical path difference up to 7 or 8 λ (λ =546 nm) as visualized by the corresponding number of interference fringes (Fig. 1). Assuming an ideal spherical shape for each region of the cell, its thickness (d) parallel to the optical axis can be calculated:
$$d = \sqrt{r^2 - (r-m)^2} \qquad (5)$$
r: radius of the sphere, m: distance of the measuring point from the outer margin as measured along r. For practical reasons only the lines of maximal and of minimal intensity in the interference fringes are used for calculations, thus determining the virtual thickness of the concentric shells by optical path differences of $\lambda/2$. These extremes may

be obtained by two alternative methods.

a) The densitogram is developed by the IBAS-system (C.Zeiss, Oberkochen F.R.G.) along a preset line (Figs. 3 and 4). The margins of the cell and the position of the extremes are indicated with a light pen on the densitogram. The order of marking the points indexes the consecutive multiples of $\lambda/2$ in optical path differences.

b) The distance of the interference fringes is determined as the distance of equidensities delineated by the image analysis system at the brightest or the darkest area (Figs. 5 and 6). Again, the differences in x-coordinates measure r and m. This procedure is limited to images with homogeneous and equal brightness of all interference fringes. In these cases no manual interaction is necessary for the determination of the parameters wanted.

Fig. 1. Interference contrast image of two <u>Psammechinus miliaris</u> egg cells about 5 min after fertilization. The fertilization membrane is clearly visible, sperm cells on the left side. In the centre of the cells the female nucleus just before fusion with sperm nucleus. Bar: 100μm

Fig. 2. Diagrammatic representation of the concentric shell model on which calculation of relative dry mass concentration is based. As an example, for calculation of c_5 at m_5, c_{1-5} are considered.

Fig. 3. Interference contrast image as viewed on the screen of an IBAS system. The line indicates the course of densimetry.

Fig. 4. Densitogram of Fig. 3

Fig. 5. Equidensities of an interference fringe pattern as shown in Fig. 3

Fig. 6. Densitogram of Fig. 5

Fig. 7. Solid concentration (ordinate) distribution from cell margin (X=0) to centre of a sea urchin egg cell (abscissa represents m-values of Fig. 2)

Solid matter concentration (c^n) at a given position (Fig. 7) is calculated according to the concentric shell

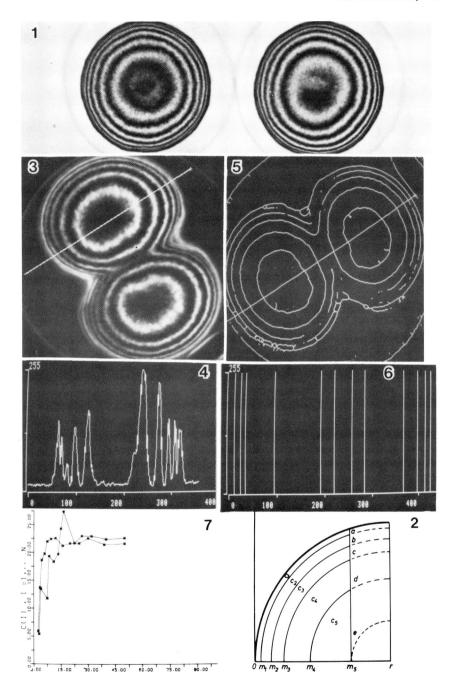

model (Fig. 2) starting from the periphery. In distances of $\Gamma = n \cdot \gamma$, m^n is evaluated by one of the methods described above and the light path through all shells with a solid matter concentration c_ν ($1 \leq \nu \leq n-1$) is considered:

$$c_n = \frac{\frac{\Gamma_n \cdot k}{\alpha} - \sum_{\nu=1}^{n-1} c_\nu \left[\sqrt{(r-m_{\nu-1})^2 - (r-m_n)^2} - \sqrt{(r-m_\nu)^2 - (r-m_n)^2} \right]}{\sqrt{(r-m_{n-1})^2 - (r-m_n)^2}}$$

Γ_n: optical path difference, k: magnification factor, α: specific refractive increment.

Figure 7 is an example of dry mass distribution in a sea urchin egg as calculated by the above procedure. Solid concentration is calculated from cell margin to center only. Thus two distribution curves are obtained for one cross section. These curves coincide in most cases, indicating a symmetrical distribution of dry mass. Temporarily, however, a transient asymmetry may arise (Fig. 7). During formation of the mitotic spindle, dry mass concentration in the egg periphery increases steadily from approximately 15% immediately after fertilization to more than 25% at the onset of cleavage furrow (Bereiter-Hahn, 1971). These observations are in good agreement with changes in the mechanical properties of the eggs (Mitchison & Swann, 1954).

DETERMINATION OF DRY MASS OF TISSUE CULTURE CELLS, THEIR NUCLEI AND NUCLEOLI

These types of measurements consititute the special domain of electronic image analyzers: Primary data are grey levels and areas. By no other method can these parameters be determined with comparable ease. To assure an optical path difference smaller than λ (Figs. 8,9), since otherwise the same grey levels would correspond to two Γ -values-differing by λ , is easily avoided by increasing the refractive index of the surrounding medium, e.g. by addition of bovine serum albumin.

a) *Experimental conditions*:

Endothelial cells (Bereiter-Hahn & Morawe, 1972) attached to a coverslip are observed while the coverslip is

Cell Interferometry / 33

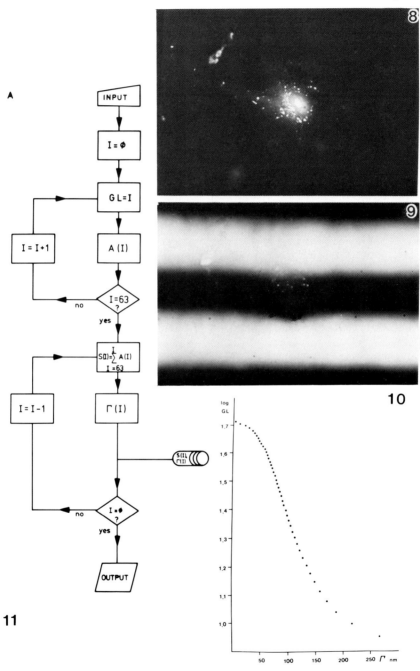

mounted either in a Dvorak-Stotler-chamber (obtained from C. Zeiss, Oberkochen F.R.G.) or in a culture chamber made from a 0.7 mm thick glass plate with a circular hole closed by two coverglasses. One without cells is used as a reference in the Mach-Zehnder-type interference microscope (E. Leitz, Wetzlar F.R.G.), equipped with a pair of 40k objectives, a stabilized Xenon lamp XBO75 and a 256nm interference filter. The cells are viewed in interference contrast. The background is adjusted to maximum darkness (first order dark) or brightness (zero-order). Analysis is done on photographic negatives (6x6, cm. Ilford FP4). Direct input of the microscope image into the image analyzer does not improve accuracy of measurements (Wientzeck et al., 1979a,b). For routine work, however, it will be more convenient. For calibration purposes interference fringes are widely separated in Figs. 8 and 9. Single XTH-cell in interference contrast (Fig. 8) and using widely separated interference fringes (Fig. 9), Optical path differences of cell and reference area in no instance exceed $\lambda/2$ (λ =546 nm)

Fig. 10. Calibration curve, where grey level, GL, (ordinate) is plotted against optical path difference, Γ (abscissa)

Fig. 11. Flow diagram of calibrating grey level as compared to extending optical path difference: A measuring horizontally that corresponds to 1 λ is overlayed on an interference fringe pattern (without cells). Number of picture points S (I), detected up to a corresponding optical path difference Γ (I) equals

$$\lambda/2 \cdot \frac{S\ \{\ I\ \}}{FRAME\ AREA} = \Gamma\ (T)$$

Γ (I) and S (I) are stored on a disc.

Equivalenly spread interference fringes are photographed at the same exposure time as the cells (Bereiter-Hahn 1977).

b) *Image analysis*:

A Quantimet 720 (Imanco, Melbourn) equipped with a plumbicon camera is used for image analysis. For limiting the field of measurement the margins of single cells or nuclei are delineated with a light pen, "image editor", on the Quantimet screen. In the following for each grey level the corresponding area in the cell is measured and used for further calculations. These are based on a calibration procedure relating differences in grey levels to optical path differences (Fig. 10.). Differences in brightness (grey level) between background and object depend on the optical path differences. These equal $\lambda/2$ (273 nm in our particular system) when a specimen appears at maximum brightness on a maximal dark background. The relation of phase differences of the interfering waves and their intensity is a $\sin^2$-function. A calibration curve like that in Fig. 10 relates each grey level difference to an optical path difference. Fig. 11 shows a flow diagram of the automatized calibration procedure.

Interference contrast images usually exhibit some degree of inhomogeneity in background brightness, which is due to the limited accuracy of dislocation of the fringes (ideal interference contrast). Electronic compensation of these inequalities in brightness distribution is not acceptable because grey levels are not related linearly to optical path differences. Therefore the grey level difference between each point of a specimen and the brightness of the surrounding cell free area has to be determined and used for the calculation of optical path difference according to the specific calibration (Fig. 12, B1,2). In practice the image is rotated by a prism until the direction of the shade coincides with the direction of the lines on the television screen of the image analyzer. The frame-slit is moved along this axis (limiting the field of measurement) step-by-step over the object. The step size equals the width of the frame. For all positions the grey level of the brightest area is determined (in this case cell free areas appear in maximum brightness) and these are used as references. Futhermore, the area covered by each of the 64 grey levels, which are discriminated by the Quantimet 720B, are recorded (Fig. 12, B3).

Fig. 12. Flow diagram of dry mass determination.
B1: Transfer of pixels to data array Z(J), control of scanning the specimen by the slit frame of a given width and the length L. Input: N (number of slits), L (slit length), B (specific refractive increment), M (magnification factor). Indices: I (slit number), Y (number of vertical blocks in the slit). B2: Determination of the brightest zone used as a reference. R: Reference grey level, J: index of pixels. B3:

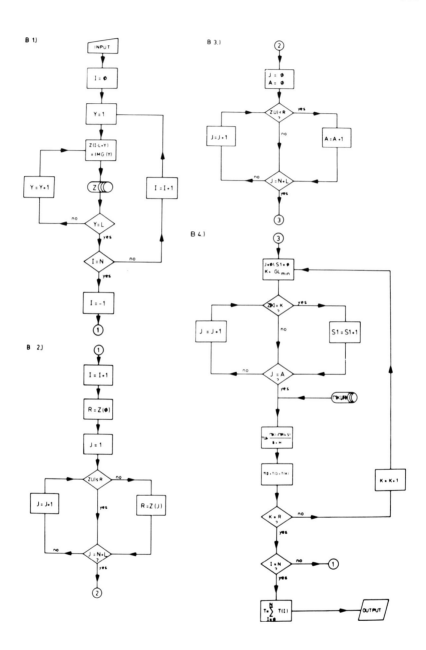

Calculation of specimen area enclosed in one slit frame. This area (A) equals the sum of all picture points darker than reference. B4: Dry mass calculation: All pixels of a certain grey level (K) are summed up [S (I)], path difference [Γ(I)] is read from disc (see diagram A, Fig. 11). T (J): dry mass calculated for each grey level. B2 and B4 are looped N times, the partial dry mass, T, (I), are summed up to partial dry masses. Output: T (total dry mass) and A (area).

The Mach-Zehnder type interference microscope allows the investigation of cells fully enclosed in an epithelial sheet, where no reference area exists. In well spread cells the very thin areas of cytoplasm will give rise to optical path differences below the limits of resolution of the system (ca λ/100), these areas are used as a base-line reference. The value is determined automatically by the analysing system in measurements of whole cells (Fig. 12, B1,2). In the case of measurements of nuclei and nucleoli the field of measurement does not include these reference areas, therefore the respective values have to be entered into the computer by hand.
C)

Dry mass calculation is based on formula (4) (see flow diagram Fig. 12, B4).

i.) For each of the 64 grey levels, which can be discriminated by the Quantimet 720 the corresponding optical path differences (Γ) to the grey level of reference regions are read in nm from the calibration table (see flow diagram, Fig. 11, 12, B1).

i.i.) The grey levels are automatically and sequentially evaluated, and the number of picture points exhibiting a particular grey level is counted ("detected points"). It measures the area (A in formula 4) which exhibits a certain optical path difference (Γ).

From the respective Γ and number of picture points partial dry mass is calculated. The sum of all the partial pixel-by-pixel determinations of dry mass of the structure under investigation, as well as the sum of all detected picture points measures the projected area of this structure. Systematic studies lead to the conclusion that reduction of lateral resolution by averaging up to 40 picture points (=1 pixel ≙10 μm^2 on the cell) does not significantly change the result of these measurements. However, it considerably accelerates the procedure of image evaluation. The actual maximum size of one pixel, for which an accurate dry mass determination can be made depends on the final magnification of a given feature on the screen and omits spatial frequency distribution. Pilot measurements of a given object using different pixel sizes are necessary to optimize the relation between lateral resolution and the time required for the measurements.

The values obtained may be stored on a disc, and used for further analysis of frequency distribution (Fig. 13), or for more extended interpretations of these data. From an extended analysis of XTH-cells we observed a steady increase of nucleolar dry mass with increasing mass which characterized two populations of cells one with nucleoli of high density and another with nuclei of lower optical density (Bereiter-Hahn et al., 1981).

As shown in formula (2) optical density is related to solid concentration. However, it cannot be calculated from dry mass and area values. As a consequence many authors have made an attempt to estimate volume. Most of the assumptions on which these calculations are based constitute very rough estimates. On the other hand, determination of dry mass per unit area also indicates solid concentration, and in this case only reliable primary data are used. In biology the determination of relationships between parameters under varying conditions are often more important than the measurement of absolute values. In those cases the representation of solid concentration in terms of dry mass per unit area suffices.

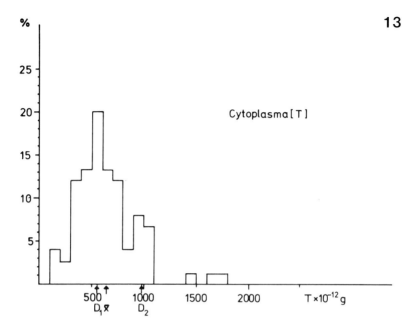

Fig. 13. Dry mass (T) distribution of primary heart endothelium cells from <u>Xenopus</u> tadpoles. The distribution curve reveals two peaks indicating two physiologically different cell populations (from Bereiter-Hahn et al., 1981).

3. DETERMINATION OF CELL VOLUME

Volume determination of irregularly shaped cells is the most difficult task that might be approached by interferometry. In principle equation (1) can be used: Γ is measured by one of the methods described, N_c and n_m have to be known, then d can be calculated. The corresponding area A is determined by image analysis as described above. The total volume is the sum of all products of the corresponding A and d values. Using interferences between light reflected from the cell surface and underlying substratum increases the accuracy of this method, because the light beam penetrates the cell twice (Bereiter-Hahn et al., 1983). This is made possible by reflection-inteference-microscopy (Beck & Bereiter-Hahn,

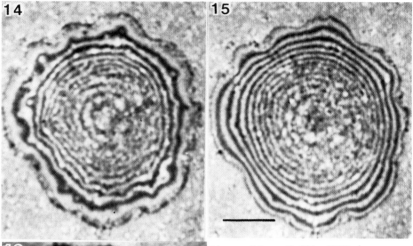

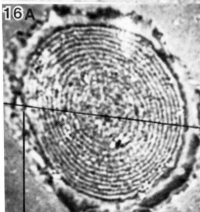

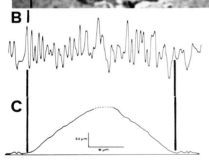

Figs. 14 and 15. RIC-image of a single Xenopus tadpole epidermis cell in normal culture medium (300 mosm) (Fig.14), and in hypotonic medium (Fig. 15) (220 mosm). Mean numerial a aperture of illumination: 0.17 Bar 10 μm. Fig. 16. Reconstruction of the cell profile (c) along the scanning line (drawn in a,) based on the densitogram shown in b. The vertical lines connect an interference ring in a with its representation in the densitogram and the related position in the profile reconstruction. The magnification scale of profile height is 5 times that of profile length. The dashed line indicates the region, where the order of interference lines cannot be determined unequivocally. Bar: 10μm

1981) as developed by Ploem (1975). Details for interpretation are given by Curtis (1964), Gingell & Todd (1979), Beck & Bereiter-Hahn (1981), and Gingell (1981).

The procedure of volume determination is performed in two steps, measuring the refractive index and measuring the optical path difference between light reflected at the cell/medium interface and at the glass/cell (or glass/culture medium) interface of cells spread on a glass surface.

Refractive index of the cells is determined by immersion refractometry (Barer & Ross, 1952, Barer & Joseph, 1954) in phase contrast. The refractive index of a balanced salt solution is varied by changing the bovine serum albumin content. Evaluations are done over a 1-2 min. interval after addition of the medium to avoid changes in cell refractive index due to pinocytotic activity. Electron microscopic studies have confirmed that pinocytosis is negligible in the epidermis cells tested, at least for a few minutes after the BSA-containing salt is added. The refractive index of the medium best matching the phase contrast appearance of the cells is determined with an Abbe refractometer. Automatic refractometry as used by Sernetz & Thaer (1970) is restricted to compact and rounded cells which exhibit bright contours when viewed in the microscope with dark field. The Leitz Diavert microscope allows an immediate change from phase contrast to reflection-interference-microscopy (RIC) with variable illuminating apertures. Illumination with a small aperture (light approximately parallel to the optical axis) emphasizes the contours of the cell/medium interface, which is visualized by the pattern of inteference fringes (Fig. 15). Those pictures are densitometrically analyzed, e.g., by the method described for analysis of the sea urchin egg. Again only maxima and minima are used for calculation of cell thickness at any given point. The distance from a dark fringe to an adjacent bright zone corresponds to a thickness difference of approximately $\lambda/4$. From the sequence of fringes, a thickness profile through any part of the cell (Fig. 16) is calculated using the equation of reflectivity at double layers of thin films as given by Beck & Bereiter-Hahn (1981).

Volume determination from thickness profiles is done by one or the other of the following methods:

a) The volumes of all thickness profiles measured are summed up. The distance of the scan lines determines the apparent thickness of each "profile disc".

b) A more accurate procedure uses the capability of image analyzers in measuring the area enclosed by each of the interference fringes. In this case the cell is supposed to represent a stack of discs; each disc, approximately $\Gamma/8$ high, corresponds to the thickness increase between a dark and a bright fringe. The exact height increase (Δd) between a bright and a dark fringe is given by

$$\Delta d = 1/8 \ (n_c \cos\gamma) - 1 \quad (7)$$

= wavelength γ : angle of incidence

Problems arise in order determination of fringes in lamella regions (Figs. 15, 16). No reliable method exists for automatic evaluation. In these cases the operator has to make judgements and to enter the data into the computer.

Measurements of alterations in cell volume in response to osmotic conditions or metabolic changes can give insights into the mechanical properties of living cells (Strohmeier & Bereiter-Hahn, in prep.).

CONCLUDING REMARKS

It now seems possible to accomplish the most tedious operations of classical microscope interferometry, the determination of area of object(s) of interest (Schiemer, 1963) and optical path difference (the product of refractive index and object thickness) by use of image analyzers. The primary data that contemporary image analyzers evaluate are brightness (grey levels) and area. Use of white light allows a rough estimate of phase differences between interfering beams by comparing the colors in the image to the range of classical colors produced by an optical prism (Gahm, 1970).

In addition it should be noted that prolonged illumination of cells with visible light will be injurious. Among the spectral lines emitted by mercury lamps, the line at 577.8 nm proved to be the least harmful to tissue culture cells (Beck & Bereiter-Hahn, 1984). Nevertheless, injuries caused by illumination can be kept to a minimum when an image intensifier camera produces the electronic image. Even if geometrical homogeneity of the image or the photometric properties of such a camera are inferior to some cameras designed especially for image analysis, intensifier cameras will be superior in studies on living cells and tend to preserve normal internal structure and reactions. Application of this method is particularly recommended for investigations with the reflection-interference-microscope.

Recent developments in the field of image analysis offer unique opportunities to follow cellular growth without interfering with cellular metabolism.

REFERENCES

Barer R. Ross KFA (1952). The refractometry of living cells. J Physiol 118:38.
Barer R, Joseph S (1954). Refractometry of living cells. Part I: Basic Principles, Quart J Micr Sci 95:399.
Barer R, Joseph S (1955). Refractometry of living cells. Part II: The immersion medium. Quart J Micr Sci 96:1.
Beck K, Bereiter-Hahn J (1981). Evaluation of reflection interference contrast microscope images of living cells. Micr Acta 84:153.
Beck K, Bereiter-Hahn J (1984). Cell damage by visible light irradiation. Europ J Cell Biol (in press).
Bereiter-Hahn J (1971). Moglichkeiten und Grenzen der Auswertung interferenzmikroskopischer Filmaufnahmen lebender Zellen. Res Film 7:302.
Bereiter-Hahn J, Morawe G (1972). Stoffwechselabhangige mitochondriale Bewegungen in endothelialen Kaulquappenherzzellen in Gewebekulturen. Cytobiol 6:447.
Bereiter-Hahn J (1977). Mikroskopische Spezialverfahren zur selektiven Kontraststeigerung fur die automatische Bildanalyse. Micr Acta Suppl 1:165.
Bereiter-Hahn J, Wientzeck C, Brohl H (1981). Interferometric studies of endothelial cells in primary culture. Histochem 73:269.

Bereiter-Hahn J, Strohmeier R, Beck K (1983). Determination of the thickness profile of cells with the reflection contrast microscope. Sci Techn. Inf. VIII:125.

Beyer H (1973). Handbuch der Mikroskopie. VEB Vlg Technik Berlin 495pp.

Curtis AS (1964). The mechanism of adhesion of cells to glass. J Cell Biol 20:199.

Gahm J (1970). The interference color chart according to Michel-Levy. Zeiss Oberkochen Werkz. 46:118.

Gingell D, Todd I (1979). Interference reflection microscopy: a quantitative theory for image interpretation and its application to cell-substratum separation measurement. Biophys J 26:507.

Gingell D (1981). The interpretation of interference-reflection images of spread cells: significant contributions from thin peripheral cytoplasm. J Cell Sci 49:237.

Mitchison JM, Swann MM (1954. The mechanical properties of the cell surface. I. The cell elastimeter. J Exp Biol 31:443.

Ploem JS (1975). Reflection-contrast microscopy as a tool for investigation of the attachment of living cells to a glass surface. In: R.von Furth (ed) Mononuclear phagocytes in immunity, infection and pathology. 405. Blackwell Oxford.

Ross KFA (1967). Phase contrast and interference microscopy for cell biologists. E Arnold Ltd London 238.

Ross KFA, Gordon RE (1982). Water in malignant tissue, measured by cell refractometry and nuclear magnetic resonance. J Micr 128:7.

Schiemer HG, Hubner K, Deseler H (1963). Zur Frage der programmierten Berechnung interferenzmikroskopischer Ergebnisse. Z wissenschftl Mikr mikr Techn 65:165.

Sernetz M, Thaer A (1970). Immersionsfraktometrie an lebenden Zellen. Z anal Chemie 252:90.

Strohmeier R, Bereiter-Hahn J (1984). Control of cell shape and locomotion by external calcium. Exp Cell Res (in press).

Wientzeck C, Brohl H, Bereiter-Hahn J (1979 a) Automatic micro-interferometry of tissue culture cells. Micr Acta Suppl 3:147 pp.

APPLICATIONS OF MICROSPECTROFLUOROMETRY TO METABOLIC
CONTROL, CELL PHYSIOLOGY AND PATHOLOGY.

Elli Kohen, Joseph G. Hirschberg, Alexander Rabinovitch

Papanicolaou Cancer Research Institute - Miami, FL 33101;
Department of Biology, Department of Physics, Laboratory
for Applied Optics and Astrophysics, University of Miami;
Division of Diabetes Research, School of Medicine
University of Miami.

INTRODUCTION

A subbranch of the optical methods devoted to the study of living cells, microspectrofluorometry (Rousseau, 1957; Olson, 1960; Caspersson et al., 1965; West, 1965) of cell coenzymes (NAD(P)H, flavins) (Chance and Thorell, 1959; Hirschberg et al., 1979; Kohen et al., 1982a,b, 1983b) in conjunction with sequential microinjections (Nastuk, 1953; Kopac, 1964) into the same cell of metabolites and modifiers, reveals aspects of the regulatory mechanisms of transient redox changes of mitochondrial and extramitochondrial dehydrogenase pathways (Kohen et al., 1982b, 1983b). A direct in situ approach aimed at the compartments and organelles of the living cell could lead to breakthroughs in unraveling the mechanisms of metabolic control and their deregulation. Concerning the accessibility of different pathways to probing by such methods, the bulk of these studies have been directed to NAD(P)H responses, but there are some limitations in the independent evaluation of different redox couples (free/bound NAD(P)-NAD(P)H) which are not necessarily in equilibrium (Krebs, 1973).

Studies involving flavins (Chance et al., 1968) are still in their infancy and present very complex problems: some flavins are not fluorescent and the participation of flavin changes in metabolic reactions is quite variable.

While a new body of biochemical knowledge, based on an

in situ approach to biochemistry with correlation to cell structure and function, can be constructed from microspectrofluorometry findings, its usefulness has sometimes been questioned. It is sometimes claimed that more extensive information may be obtained from studies with isolated enzymes and cell fractions, with the optimistic view that what may be learned from the parts can be put together to understand the function of the whole. The main strength and contribution of microspectrofluorometry, however, is not so much in the description of NAD(P)H or flavin pathways themselves, but rather in the ability to detect how these same pathways behave at a given cell site, compartment or organelle under the dynamic conditions of cellular activity; i.e. when the cell, be it normal or pathological, is behaving and functioning as a living entity. Concerning this aspect the other methods, while yielding more details in the evaluation of a particular enzyme or pathway, are totally powerless.

NAD(P)H responses to sequential injections into the same cell of glycolytic, respiratory or shunt substrates and modifier adenine nucleotides and ions, disclose the intracellular regulation and compartmentation of bioenergetic pathways (Kohen, et al., 1983b). Multicellular integrated states and cell-to-cell inhomogeneities are observed which can be reconciled with different patterns of cell-to-cell metabolic coordination or its failure (Kohen et al., 1979, 1983a; Thorell, 1978). Thus, coenzyme-based microfluorometry-microinjection studies are relevant to long standing questions concerning the distribution of control over enzymes participating in a metabolic system, or the coherent/incoherent behavior of multicellular systems.

While the arguments above concern the function of mitochondrial, cytosolic and nuclear dehydrogenase pathways, other cellular organelles, sites and functions are amenable to exploration by exogenous fluorescent or fluorogenic probes (Bereiter-Hahn, 1976), fluorescent photosensitizers (Scott et al., 1976), carcinogens (Gelboin, 1980), and xenobiotics. The dynamic functional mapping of the living cell in terms of microcompartments, activity of cell organelles and xenobiotic-metabolizing systems becomes a real possibility, based on uni/two-dimensional topographic fluorescence scans and spectral analysis of emissions at different cell sites.

METABOLIC CONTROL

Two alternative approaches are available:

1) first the injection of a modifier in order to bring all the enzymes in the probed metabolic system to the desired metabolic condition and then the injection of a substrate;

2) first the injection of a substrate to induce a transient NAD(P)H fluorescence, followed by the injection of a modifier at selected times with resulting alterations of the transient course.

This second approach, while yielding key information, suffers from a drawback since the enzymes of the probed system will not be 'tuned' when the substrate is injected and therefore the effect of the modifier on all sorts of NAD(P)H responses will be observed, complicating the interpretation.

The modeling of metabolite responses and modifier effects helps to verify kinetic and irreversible thermodynamic hypotheses referring to the regulation of metabolic systems (Westerhoff et al., 1981; Westerhoff, 1983). The injection of ADP in the course of an NAD(P)H transient produced by glycolytic (e.g. glucose-6-P = G6P) or respiratory substrate (e.g. malate) leads to a sharp reoxidation analogous but not identical to the State III of isolated mitochondria (Chance and Williams, 1955), followed by a spontaneous resumption of NAD(P) reduction (when ADP is exhausted) analogous to a State III to IV transition and ultimately a return to the initial redox state. The response to ADP alone is biphasic, i.e. an initial acceleration of oxidative phosphorylation causing oxidation of NADH, with a subsequent large NADH transient increase due to coenzyme reduction. The sequence of oxidation-reduction cycles may be explained as above in the case of the ADP-interrupted transient, but this time based on changes of endogenous substrate shifts and allosteric effects on mitochondrial-extramitochondrial dehydrogenases rather than injected substrate.

The responses of whole metabolic systems contrasted to those of single enzymes lead to a new terminology (Kacser, Burn 1979) with definitions of coefficients and parameters,

taking into account the distribution of control over the
enzymes participating in the system, which are extremely
sensitive to changes in the exact concentrations of the
metabolites and modifiers present.

METABOLIC COMPARTMENTATION

From a theoretical and functional standpoint the identification of metabolic compartments in the intact living cell introduces in biochemistry the concept of structure and function at a higher level of organization than heretofore, such as multienzyme complexes and aggregates bound to intracellular membranes or segregated by them, i.e. as within organelles (Chance, 1963; Welch, 1977; Moses, 1978; Srere, 1978). Two-step NAD(P)H responses to sequentially injected glycolitic (G6P) and respiratory substrates (malate), identify the mitochondrial-extramitochondrial compartments.

Further metabolic and structural compartments are unraveled, e.g. in L cells (Sanford et al., 1948), L cells grown in hypertonic media (LS cells) (Schachtchabel et al., 1972; Clegg, 1979; Mansell and Clegg, 1983), and poorly differentiated rhabdomyosarcoma cells (McAllister et al., 1969). A two-step NAD(P)H response to sequentially injected fructose-1,6-diphosphate and G6P indicates the dynamic or even structural compartmentation of glycolytic phosphate esters in separate intracellular pools (Moses, 1978). Especially in the case of small foci of increased NAD(P)H fluorescence, changes with an intensity barely above noise level are observed within 0.1 sec of the glucose-1-P (G1P) injection. Within another 0.1 sec the response is further generalized but still focal. It may take 4-20 sec to observe the maximal and fully generalized response.

The multifocal response to G1P is at times observed to start at a nuclear site, an observation more evident when the shunt substrate 6-phosphogluconate (6-PG) is injected into LS cells. The reductive response to 6GP appears in topographic fluorescence curves fairly circumscribed by the boundaries of the nucleus (Figs. 1 & 2). Such localization of the shunt pathway seems compatible with the NADPH requirement for nucleic acid synthesis. The focal and generalized responses are not inconsistent with a structural microarrangement based on enzyme activity bound to intracellular surfaces and generalization of the response

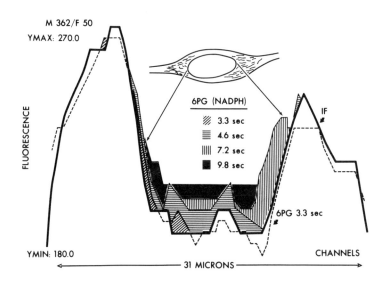

Fig. 1
Topographic plots of the NADPH response to injection of 6-phosphogluconate (6PG) in a rhabdomyosarcoma CCL136 cell. The ordinate shows the lower (YMIN) and upper (YMAX) limits of the fluorescence intensity measured in counts per single scan time (i.e. counts per $\sim$ 0.1 sec); the abscissa shows the distance in micrometers along the scan line viewed by the unidimensional multichannel detector. For these and following computer traced plots the experimental log number is identified on the left upper corner (e.g. M362/F50). Each channel corresponds to a cell width of 0.6 micrometers. The cell scanned is drawn above with corresponding parts of the topographic plots (i.e. nuclear boundaries) indicated by arrows. IF = initial fluorescence plot prior to microinjection of 6PG. The fluorescence increase (focal first, then generalized) due to NADPH, as seen in sequential plots (at 3.3, 4.6, 7.2 and 9.8 sec after injection) is largely localized within the boundaries of the nuclear region. Thus, consistent with cytological studies revealing the localization of glucose-6-P dehydrogenase at the nuclear membrane, another enzyme of the hexose monophosphate shunt pathway, i.e. 6PG dehydrogenase, also seems associated with the nucleus.

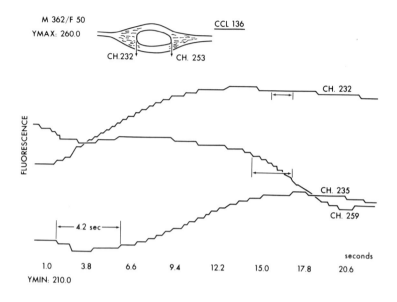

Fig. 2
Time plots of the NADPH response to 6PG obtained from different sites of the same cell as in Fig. 1. Ordinate: fluorescence intensity (definitions of YMIN, YMAX are the same as in Fig. 1); abscissa; time after injection. The cell is drawn above with positions viewed by detector channels 232 and 253 indicated. Injection of 6PG takes place slightly before sec 1. An increased fluorescence (NADPH) is observed following the injection in channels 232, 235 within or on the edge of the nucleus. There is a decrease in channel 259 outside the nucleus.

---

through changes of the adenylate ratio in the bulk aqueous phase (cytosol) of the cell (Goldbeter and Lefever, 1972; Welch, 1977). In Figures 1 and 2 peripheral cytoplasmic regions show a paradoxical oxidation. In living cells the oxidoredox response to injected substrate is known to be bistable (Hess, 1973); depending upon local inhomogeneities in the levels of metabolic intermediates, an oxidative or reductive response is observed with the same substrate.

Observation of multifocal responses to respiratory substrates is facilitated by early kinetic studies and growth conditions (<u>e.g.</u> high ionic strength) conducive to high density and enhanced functional activity of mitochondria. Distinct extramitochondrial malate dehydrogenase (MDH) and mitochondrial MDH compartments (Veech et al., 1969) are identified in high NaCl-adapted LS cells (Fig. 3). Thus, upon injection of malate a NAD(P)H response is first detected in a mitochondria-poor region of the cell (often coincident with cytoplasm overlying or underlying the nucleus) to be followed 1 to 7 sec later by a response at the site of perinuclear mitochondrial aggregates. In LS cells, two asynchronous components (cytosolic and mitochondrial) of the malate response are detected, while the response to isocitrate is nearly confined to the site of the delayed malate response (mitochondrial) (Fig. 4).

THE ROLE OF THE ADENYLATE CHARGE IN THE LOCALIZATION, GENERALIZATION AND INTENSIFICATION OF THE RESPONSE TO SUBSTRATE

Generalization of the metabolic response to injected substrates (e.g. respiratory) is accelerated with concomitant enhancement of the response upon:

1) pretreatment with an ATP trap (ethionine) (Faber, 1963), Fig. 5.

2) pretreatment with an uncoupler (dinitrophenol) Fig. 6.

Since both conditions result in ATP depletion, they point to the role of the adenylate charge in metabolic regulation. A similar but weaker intensification of the metabolic response to injected substrate is observed in cells submitted to acute hypertonic shock, a condition which may also lead to ATP depletion.

Analogies between the behavior of MDH in LS cells treated with the ATP trap ethionine and the carcinogen dimethylnitrosamine (DMN) (Magee and Barnes, 1967) suggest a similar mechanism. The ultimate surge of a strong response to respiratory substrates in DMN-treated cells was attributed to exhaustion of the shunt pathway as a result of the heavy demand on NADPH for carcinogen metabolization. Concomitant depletion of ATP stores with accomplishment of a specific function (carcinogen or xenobiotic detoxification or metabolization) is conceivable.

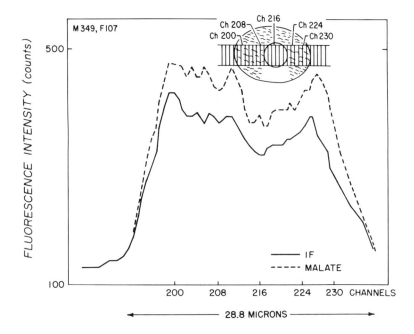

Fig. 3A
Topographic plot of the response to malate in an L cell grown in a hypertonic NaCl medium (0.225 M). Abscissa and ordinate are the same as in Fig. 1. An ATP trap, ethionine (6 mM) was added to facilitate the surge of a large NAD(P)H response to malate. The injected cell is drawn above to illustrate the presence throughout the cell of closely packed long filamentous mitochondria. Two consecutive plots are drawn: IF (initial fluorescence) and the level of NAD(P)H about 20 seconds after injection of malate. The central depression corresponds to the nucleus and the two lateral prominences to regions densely packed with mitochondria. The increase of fluorescence appears more pronounced at the sites of these prominences.

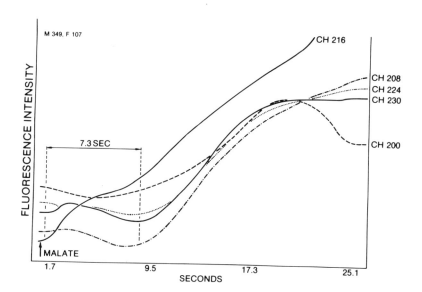

Fig. 3B
Time curves of NAD(P)H responses to malate for the same cell as in A, recorded for five different sites (regions viewed by detector = channels (CH) 200, 208, 216, 224, and 230). The scale of 28.8 microns in Fig. 3A corresponds to the interval between channels 192 and 240. Abscissa and ordinates are the same as in Fig. 2. The longest delay in the response to malate is observed between channel 216 (corresponding in Fig. 3A to the central depression between left and right prominences, i.e. the site of the nucleus with underlying and overlying cytoplasm which contains rare mitochondria), and channel 200 (left prominence, in Fig. 3A a site of high mitochondrial density). The malate-induced fluorescence rise starts at site 216, about 7.3 seconds earlier than at site 200. This delay could represent the time interval between the responses of extramitochondrial and mitochondrial malate dehydrogenases, i.e. the delay at the mitochondrial membrane barrier. Within the framework of metabolic studies in the living cell, this is the best documented example so far of compartmentation and metabolic lag due to an intracellular membrane barrier.

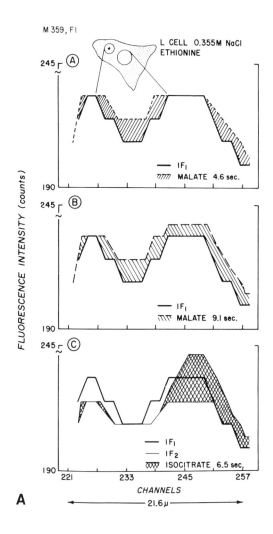

Fig. 4A

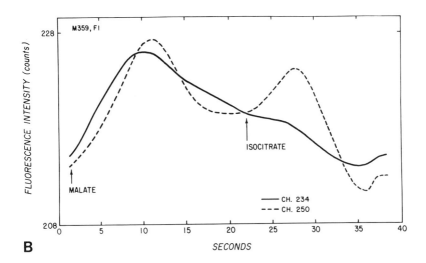

Fig. 4
Topographic (A) and time (B) curves of the responses to malate and isocitrate sequentially injected into a binucleated L cell grown in the presence of 0.355 M NaCL.
A: Three consecutive metabolic responses:
A.A: Malate response at 4.6 sec (probably extramitochondrial); IF_1 = initial fluorescence before injection of malate. Arrows point to the approximate limits of the cell region covered by the two nuclei as projected on the topographic plots.
A.B: Delayed malate response at 9.1 sec (most likely mitochondrial) followed by reoxidation of NAD(P)H (see curve IF_2 in A.C.
A.C: Response to isocitrate largely coincident with delayed malate response.
B: Time curve from a site showing the early (probably extramitochondrial) malate response in Fig. 4A (channel 234) and a site showing the delayed (probably mitochondrial) malate response (channel 250). The maximum of the malate response is indeed delayed by a few sec in channel 250 compared to 234. The malate response is reductive at both sites (channels 234 and 250), but a reductive response to isocitrate is only observed at the presumably mitochondrial site (channel 250).

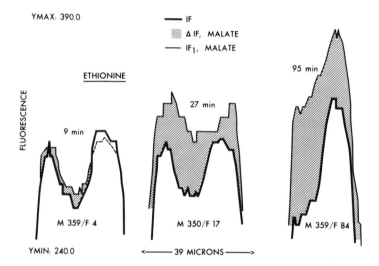

Fig. 5
Topographic plots of the malate responses in three L cells grown in 0.355 M NaCl and preincubated for different lengths of time in the presence of the ATP trap ethionine (6 mM). Abscissa and ordinate, the same as in Fig. 1, but for three cells at indicated times after addition of ethionine.
IF = initial fluorescence prior to injection of malate;
Δ IF = fluorescence increase due to injection of malate;
IF_1 = fluorescence intensity of NAD(P)H after injection of malate.
At 9 minutes after addition of ethionine the metabolic response is quite weak, and comparable to that seen in similar cells untreated with ethionine. A considerable enhancement of the metabolic response is observed at 27 and 95 minutes, compared to 9 minutes.

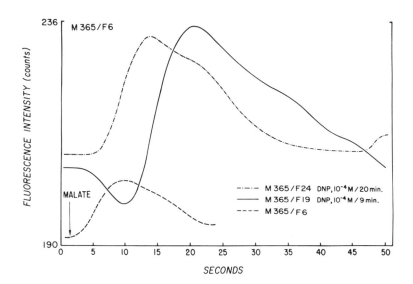

Fig. 6
Time plots of the malate responses in three L cells grown in the presence of 0.355 M NaCl.
Cell M365/F6 control; cells M365/F19 and F24, respectively, 9 and 20 minutes after addition of 10^{-4} M dinitrophenol (DNP).
Abscissa and ordinate are the same as in Fig. 1.
A considerable enhancement of the metabolic response is seen after addition of DNP.

RECOGNITION OF QUALITATIVE/QUANTITATIVE METABOLIC DIFFERENCES IN NORMAL VERSUS PATHOLOGICAL CELLS

In studying the bioenergetics of living cells the glycolytic/respiratory ratio and the response to ATP depletion provide preliminary information on the characterization of cells. Responses are compared in L cells, the same adapted to hypertonic media (LS) and highly malignant rhabdomyosarcoma CCL 136 cells. The largest responses to respiratory and lowest response to glycolytic substrates are in the LS cells. The converse is observed in the CCL 136 where there is no lack of mitochondria, but they may be functionally deficient, as suggested by the predominant response to G6P compared to malate. In the control L cell the malate and G6P responses are relatively well balanced. Pretreatment with an ATP trap leads to dramatic intensification of the NAD(P)H response to injected malate in the LS cells, but the effect is weaker in CCL 136 with a minimal response to malate and some intensification of the glycolytic response.

No attempt was made to determine the grade of malignancy in these cells. However, the extensive mitochondrial rearrangement in the LS cells, accompanied as it is by high functional activity, seems strongly reminiscent of the seagull salt gland cells which function as a miniature desalinization plant. Conversely the mitochondria of the CCL 136 appear functionally deficient. Thus in a functional sense, of the three cells considered, LS cells are the closest to a specialized cell and the CCL 136 at the other end exhibit poor differentiation. Inferences from these different responses of respiratory/glycolytic pathways to ATP depletion suggest graded differences in the ATP synthesis/hydrolysis cycle. From the response to ATP traps in the LS cells, it may be inferred that prior to treatment with the trap their ATP stores are not depleted despite high energy requirements for the ion pump. This suggests a favorable stoichiometry in terms of ATP molecules hydrolyzed per ion pump. The same cannot be said of the CCL 136 in which the response to the trap is minimal.

While in an initial phase, carcinogen-treated or early malignant cells maintain a positive economy of ATP, poorly differentiated Ehrlich ascites cells are not as sensitive to ATP traps. ATPase activity related to the ion pump seems to function less efficiently. However, if these cells are adapted to high NaCl concentrations, their metabolic activity and responses to ATP traps return to a state suggesting a more efficient functioning of pump-related ATPase activity (Racker, 1977).

THE IDENTIFICATION OF FREE AND BOUND COENZYMES

The identification of responses associated with free and bound coenzymes can be of crucial significance in terms of cell type, metabolic state or adaptive changes. There are allegations that loss of binding sites in malignant cells leads to changes in the bound/free coenzyme ratio.

The sum emission of living cells under excitation at 365 nm is due to bound and free NAD(P)H plus a small contribution from flavins. Resolution was first attempted in the simplest case, i.e. the increase spectrum (ΔI_F) due to microinjection of G6P (Salmon et al., 1978). Weber (1961) developed a method of matrix analysis to determine the number of individual fluorescing species in a sum emission. The intensity of emission at different wavelengths is determined for fluorescence excited at different wavelengths and matrices of these intensities are formed. The rank of these matrices gives the number of fluorescent emitters. For example, if the 2 x 2 matrices are significant there are at least two emitters, if the 3 x 3 are significant there are at least three emitters, and so on.

Resolution of NAD(P)H spectra in ascites cells microinjected with G6P reveal two components in the increase spectrum due to NAD(P) reduction by substrate (free and bound NAD(P)H), and three components in the initial spectrum (I_F) prior to injection (free, bound NAD(P)H and flavins). Thus, when a 3 x 3 matrix is constructed for the I_F emission, the determinant of the matrix is found to be significant compared to a norm of the matrix and the noise of the signal, indicating the significance of the matrix.

The participation coefficients of the three components are seen to vary with aerobic-anaerobic transition, treatment with inhibitors (oligomycin) and differentiated state of the cells (e.g. poorly differentiated vs. well differentiated melanoma cells).

MULTICELLULAR INTEGRATED STATES, CELL-TO-CELL METABOLIC INHOMOGENEITIES AND MODULATION OF INTERCELLULAR FLOW

The delimitation of functionally integrated microdomains of communicating homologous and/or heterologous cells (Loewenstein and Kanno, 1964) can be achieved by the injection of tracers (fluorescein, lucifer yellow, Stewart, 1978) or metabolites and modifiers. Metabolic fluxes at any point within a cluster of cells proceed in a spatiotemporal structure. A unifying hypothesis among cytologists and enzymologists is that the extensive membrane compartmentalization in the cytoplasm and related structural arrangment is to serve as a matrix upon which enzymes can be assembled into ordered multienzyme units to direct the flow of metabolites, protons, electrons, etc (Welch, 1977).

In monolayer cultures of pancreatic islet cells, small territories of 2-10 islet cells appear functionally integrated by junctional intercellular communication. Communication between insulin-containing B cells and glucagon-containing A cells or somatostatin-containing D-cells has been detected (Meda et al., 1982). Cell communication was significantly increased by high glucose concentrations, the phosphodiesterase inhibitor 3-isobutyl-1-methylxanthine and the calcium ionophore A23187 (Kohen et al., 1983a). In the presence of somatostatin, cell communication is significantly inhibited. Parallel changes in hormone secretion suggested that cell communication may have participated in the regulation of islet cell secretory activity.

Microfluorometric studies aimed at examining cell-to-cell metabolic cooperation (Fig. 7) provide new evidence for establishing a link between the glycolytic cascade, the "recruiting" of islet cells and insulin secretory activity. It has been hypothesized that insulin release is secondary to the increase in glycolytic and oxidative fluxes evoked in the islet cells by a rise in the extracellular concentration of glucose (Malaisse et al., 1982). Such fluxes and the recruiting of adjacent cells in cooperative metabolic activity can be evaluated in the presence of various activators/inhibitors of insulin secretion.

Parameters derived from microfluorometry are:
1) Intercellular transit times
2) Metabolic lags in injected cells and neighbors
3) Coenzyme reduction and reoxidation fluxes

A cascade-like acceleration of glycolysis by the potent activators of phosphoglucomutase and phosphofructokinase, i.e. by fructose-2,6-P and glucose-1,6-P, has been postulated for the recruiting of islet cells in insulin secretion. Using a two dimensional multichannel microfluorometer to scan homologous/heterologous cells within a cluster, crucial hypotheses concerning metabolic events associated with hormone secretion could be evaluated in the presence of secretagogues or modulators of hormone secretion.

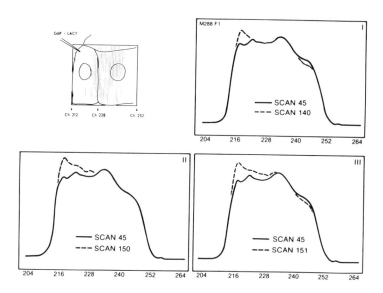

Fig. 7A
Topographic analysis of metabolic communication in islet cells from a monolayer culture of neonatal rat pancreas, as a function of time.
The insert at the upper left shows the injected cell and adjacent cell as scanned topographically, superimposed on the viewing channels of the fluorescence detector.
Frames I, II, III show the time course of topographic scans across two adjacent pancreatic islet cells. The abscissa shows channels across the two cells along the linear scan axis. Each channel corresponds to about 0.6 micrometer of cell width. The ordinate shows NAD(P)H fluorescence intensity. Scan 45 (4.5 sec), which is shown in all the frames, is recorded before the injection of substrate. Scans 140 (14.0 sec), 150 (15 sec), 151 (15.1 sec) in Frames I, II and III, respectively, are recorded after the injection of substrate (glucose-6-P (2 parts) + lactate (1 part)). Lactate was added to slow down NAD(P)H reoxidation and facilitate detection of the NAD(P)H transient.
The time interval between two consecutive scans is about 0.1 sec.

Initially the fluorescence increase due to substrate
injection is localized on the channels corresponding to the
injected cell (see scan 140, Frame I) and appears as a
prominence on the left; the intercellular region between
the injected cell and its neighbor is represented by the
trough between the left prominence and the right
prominence (which corresponds to the as yet non-responding
neighbor of injected cell). Scans 150 and 151 in Frames II
and III show that as the fluorescence response increases it
spreads gradually towards the channels corresponding to the
non-injected adjacent neighbor.

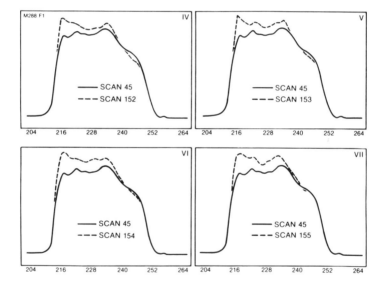

Fig. 7B
Scans 152-155 in Frames IV to VII show that the
fluorescence response corresponding to NAD(P)H has extended
to both cells.

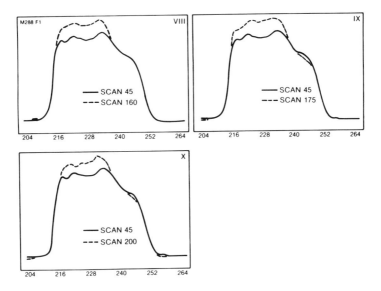

Fig. 7C
Frames VIII and IX show the maximal NAD(P)H response over both cells, with a slight regression of the response in Frame X. The time interval from Frame I in Fig. 1A to Frame X in Fig. 1B is less than 2 sec.
The sequence frames I-X represents <u>what we believe to be the first instance of detailed topographic/temporal analysis of cell-to-cell metabolite transfer identified by the NAD(P)H response in pancreatic islet cells.</u>

It should be noted that:
1) the probe for metabolic communication (NAD(P)H) is not the molecule(s) actually moving cell-to-cell
2) Studies are two-dimensional while in the actual endocrine tissue three-dimensional communication and cell recruiting are probable
3) Non-homologous (e.g. endocrine and fibroblastoid) cells which establish communication in monolayer cultures may not do so in the actual tissue

4) Direct correlation between metabolic cooperation in monolayer cultures and junctional permeability is not feasible because the freeze-fracture method required for evaluating junctions is difficult to apply here

5) Low-glucose conditions under which metabolic cooperation is observed are non-physiological for insulin secretion studies.

While the intercellular transfer of NAD(P)H, a molecule of over 700 daltons, is minimal in islet cells, this coenzyme is a suitable indicator of metabolic responses due to cell-to-cell movement of injected glycolytic intermediates and their products along the glycolytic chain (molecules in the range of 150-300 daltons). Three-dimensional studies of cell communication in isolated whole islets of Langerhans have been achieved by Michaels and Sheridan (1981). While the application of the freeze-fracture method for monolayer cultures presents difficulties, parallel studies are possible on pellets of islet cells maintained under identical conditions to the islet cells in monolayer cultures used for fluorescence studies. Non-physiological low extracellular glucose concentrations were used in early studies of metabolic cooperation because, due to bistable metabolic reactions, the direction of coenzyme fluorescence transients is reversed (i.e. oxidation instead of reduction) in cells incubated at glucose concentrations as low as 1 mM. NAD(P)H transients may be observed in the presence of more realistic higher glucose levels by the tuning of ATP levels. Studies with flavin changes also hold promise in the evaluation of metabolic cooperation in islet cells.

Different hypotheses related to hormone secretory mechanisms or pathogenesis of endocrine disease at the cellular level are amenable to exploration. Hypotheses of insulin release mechanisms address two possibilities upon stimulation by an increase in the extracellular glucose: 1) increases in the availability of glycolytic intermediates (substrate-site) and 2) increases in the activity of the key glycolytic enzyme phosphofructokinase PFK (regulatory site). PFK is a master enzyme of glycolysis (Boiteux et al., 1980) and its activation can be measured from differences in transient lag and rise kinetics for substrates proximal and distal to PFK.

Metabolic states may also differ in adjacent cells,

thereby producing a mosaic functional pattern in groups of cells. The occurrence of oxidizing or reducing responses to injected substrates is dependent on the intracellular environment (e.g. levels of different intermediates). Thus, if two adjacent cells are not in the same metabolic condition, even if they are immersed in the same extracellular medium, their metabolic responses may vary.

Whether multicellular integrated assemblies are confined to microdomains or whether they consist of a mosaic of cells exhibiting bidirectional redox changes (Hess, 1973), there seems to be a pattern, at least for islet and hepatoma cells, of "fine regulation" rather than massive communication through broad networks of cells. Other patterns of communication observed, such as delayed and remote communication to distant neighbors via a tentacle-like cell process, could be the means of integrating metabolic processes between different microdomains. Such remote communication is not especially frequent between islet cells, and most islet territories are separated from each other. Key questions are raised as to metabolic cooperation in cultures and co-cultures of differentiating cells, and possible coupling between induced and target cells. Evidence has been obtained that signal molecules between homotypic and heterotypic cells may include metabolites, modifiers, ions and possibly hormones acting on adjacent cells in a paracrine fashion.

Highly undifferentiated malignant cells such as ascites cells do not communicate. There are graded differences in communication between well and poorly differentiated melanoma cells or glia versus glioma cells. Ordinarily non-communicant L cells are occasionally seen to communicate massively in highly crowded cultures showing contact inhibition, and also in the presence of dibutyryl cyclic AMP (Azarnia et al., 1981). In view of observed graded differences according to cellular differentiation or growth patterns, determinations of intercellular transit times or flow kinetics are valuable. Thus, in L cells showing contact inhibition, intercellular transit times are at the bottom of the scale compared to highly communicant liver cells, normal fibroblasts and islet cells.

INTRACELLULAR INTERACTIONS OF FLUORESCENT CARCINOGENS

The changes undergone by the fluorescent carcinogen benzo(a)pyrene (BP) in malignant L cells (Salmon et al., 1978), inducible Buffalo rat liver (BRL) cells (Whitlock et al., 1974a,b) and oncogenic $C_3H/10$ T 1/2, Cl 8 clone (CCL226) cells (Heidelberger, 1975) have been followed. Since BP is converted to phenols, epoxides, quinones, dihydrodiols, diol epoxides and water soluble conjugates (Gelboin et al., 1980), the interpretation of blue- and red spectral shifts presents considerable difficulty but certain facts emerge (Kohen et al., 1983c). The sequence of blue-red shifts expressive of intracellular interactions and detoxification of the carcinogen is accelerated in the induced BRL compared to non-induced, as well as in the malignant and inducible lines compared to the oncogenic line. Results in carcinogen-adapted malignant lines raise intriguing questions concerning the mechanisms associated with resistance to carcinogens, activation of metabolization pathways, and progression beyond toxic products to detoxification products readily eliminated. If more rapid progression towards detoxification would denote better protection from or resistance to the carcinogen, then the malignant L cell and the induced BRL are more resistant than the non-induced BRL and the non-malignant CRL.

The predominance of blue-shifted emitters in the nucleus of L and induced BRL cells reminds one of the "Tunneling Hypothesis" (Birks, 1959; Mason, 1959) according to which blue-shifted compounds with higher energy levels of their electronically excited states will be more aggressive (reactive) for macromolecules such as nucleoproteins or even organelle membrane proteins (in mitochondria and lysosomes). The predominance of a red emitter in the most oncogenic line (CRL 226) and of a blue emitter in the non-oncogenic BRL seems in apparent contradiction to tunneling, but there are interpretations not necessarily exclusive of this hypothesis: the blue shift of the BRL may be due to a relatively harmless conjugate rather than a diol epoxide considered as an "ultimate carcinogen" (Gelboin 1980). Thus, while BP glucuronopyranoside is red shifted compared to the non-metabolized BP, BP sulfate emission is coincident with the blue shifted emission of BRL cells.

The most toxic ultimate carcinogens do not need to be present in high quantities and their emission may be overwhelmed by that of less toxic metabolites or detoxification products, which would mean that microspectrofluorometric analysis may be geared to the detection of detoxification rather than toxic products. In order to detect a very active but small component reponsible for the toxic effects, a combination of highly sensitive techniques may be required, i.e.:

1) topographic scan of fluorescence spectra from minute regions of the cell;

2) application of matrix analysis (Weber, 1961) in conjunction or not with phase selective fluorescence quenchers for resolution of spectra;

3) higher resolution spectra;

4) comparison with a catalog of known fluorescent metabolites, conjugates or adducts.

We are encountering, in the living cell, a variety of molecules which are present at cytoplasmic and nuclear locations. They could interact as well with cell organelles and organelle membranes, the microarchitectural networks of the intact cell (Clegg, 1979) as they might with nucleic acids and proteins. Studies with non-fluorescent carcinogens reveal profound alterations of mitochondrial/extramitochondrial bioenergetic pathways from the first minutes of carcinogen addition, and many of these effects are extranuclear. Better identification of intracellular carcinogen products should proceed in parallel with in situ exploration of their action at organelle sites within and outside the cell nucleus.

PHOTOBIOLOGICAL STUDIES WITH FLUORESCENT PHOTOSENSITIZERS

UV irradiation of cells treated with furocoumarins allows the detection of fluorescent monoadducts and the kinetic study of their conversion to non-fluorescent diadducts (Scott et al., 1976; Johnston et al., 1977; Moreno et al., 1981). These studies are complicated by the production of highly fluorescent endogenous photoproducts spectrally overlapping with monoadduct emission. Monoadducts are identified not only within the nucleus, but also at cytoplasmic and nucleolar sites compatible with binding to RNA as well as DNA. The binding of furocoumarins to cytoplasmic organelles and membranes as well as DNA is significant from theoretical and therapeutic standpoints.

CONCLUSION

The microspectrofluorometric approach using coenzyme reactions was initiated to comprehend, on a basic level, metabolic control and compartmentation. However, the development of a biochemical study based on living cells may gradually lead to a better understanding of cell physiology and cell pathology as well as regulatory processes or dysfunctions affecting multicellular processes. Applications to a broad variety of biomedical subjects becomes possible, including problems related to metabolic or endocrine disturbances, carcinogenesis, hereditary diseases and aging.

ACKNOWLEDGMENTS

This work was supported by National Science Foundation Grants PCM 7708549,02-03, PCM-8303691, HEW National Cancer Institute Grant CA 21153-05,06, HEW Institute of Arthritis, Metabolic and Digestive Diseases Grant AM 21330-04,06.

The authors acknowledge thankfully the stimulating comments of Dr. Britton Chance, Johnson Research Foundation, University of Pennsylvania, on the metabolic aspects of the described research, particularly in what refers to mitochondrial-cytosolic relationships.

The authors are thankful to Dr. Zbynek Brada, Papanicolaou Cancer Research Institute, for useful advice and discussions on studies with L cells and CCL 136 rhabdomyosarcoma cells maintained in high NaCl media or treated with ATP traps.

The authors are also indebted to Dr. Alain W. Wouters, Plasma Physics Laboratory, Princeton University, for advice on microspectrofluorometry of carcinogens. A posthumous tribute is owed to the late Professor Bo Thorell who inspired the initiation of the reported studies on metabolic control and compartmentation.

REFERENCES

Azarnia R, Dahl G, Loewenstein WR (1981). Cell junction and cyclic AMP: III Promotion of junctional membrane permeability and junctional membrane particles in a junction-deficient cell type. J. Membrane Biol. 63:135.

Boiteux A, Hess B, Sel'kov E (1980). Creative functions of instability and oscillations in metabolic systems. Curr Top Cell Regul 17:171.

Bereiter Hahn J (1976) Dimethylaminostyrylmethylpyridiniumiodine (DASPMI) as a fluorescent probe of mitochondria in situ. Biochim Biophys Acta 423:1.

Birks JB (1959) Change transfer complexes in biological systems. Disc. Faraday Soc 27:243.

Caspersson T, Lomakka G, Rigler Jr R (1965) Registrierender fluoreszenzmikrospektrograph zur bestimmung der primar-und sekundar fluoreszenz verschiedener zellsubstanzen. Acta Histo Chem Suppl 6:123.

Chance B (1963) Localization of intracellular and intra mitochondrial compartments. Ann N Y Acad Sci 108:322

Chance B, Mela L, Wong D (1968) Flavoproteins of the respiratory chain, in Yagi K (ed): "Flavins and Flavoproteins," Baltimore University Park Press and Tokyo: Univ of Tokyo Press, p 107.

Chance B, Thorell B (1959) Localization and kinetics of reduced pyridine nucleotide in living cells by microfluorometry. J Biol Chem 234:3044.

Chance B, Williams GR (1955) Respiratory enzymes in oxidative phosphorylation I. Kinetics of oxygen utilization. J Biol Chem 217:383.

Clegg, JS (1979) Metabolism and intracellular environment: the vicinal-water network model. In Drost-Hansen W, Clegg J (eds) "Cell Associated Water," New York: Academic Press, p 363.

Faber E (1963). Ethionine carcinogenesis. Advan Cancer Res 7:383.

Gelboin HV (1980) Benzo(a)pyrene metabolism, activation and carcinogenesis; role and regulation of mixed function oxidases and related enzymes. Physiol Rev 60:1107.

Goldbeter A, Lefever R (1972) Dissipative structures for an allosteric model: applications to glycolytic oscillations. Biophysical J 12:1302

Heidelberger C (1975). Chemical carcinogenesis. Ann Rev Biochem 44:79.

Hess B (1973). Organization of glycolysis. Symp Soc Exp Biol 27:105.

Hirschberg JG, Wouters AW, Kohen E, Kohen C, Thorell B, Eisenberg B, Salmon JM, Ploen JS (1979). A high resolution grating microspectrofluorometer with topographic option for studies in living cells. In Talmi Y (ed) "Multichannel Image Detectors," Washington, American Chemical Society, p 263.

Johnston BH, Johnson MA, Moore CB, Hearst JE (1977). Psoralen-DNA photoreaction: controlled production of mono- and diadducts with nanosecond ultraviolet laser pulses. Science 197:906.

Kacser H, Burn JA (1979). Molecular democracy: who shares the controls? (1979). Biochem Soc Trans 7:1149.

Kohen E, Kohen C, Thorell B, Mintz DH, Rabinovitch A (1979). Intercellular communication in pancreatic islet monolayer cultures; a microfluorometric study. Science 204:862.

Kohen E, Kohen C, Hirschberg JG, Wouters A, Bartick PR, Thorell B, Bereiter-Hahn J, Meda A, Rabinovitch A, Mintz D, Ploen JS (1982a). Examination of single cells by microspectrophotometry and microfluorometry. in Baker P.F. (ed) "Techniques in Cellular Physiology-Part I," Amsterdam, Elsevier/North Holland Scientific Publishers, Ltd, p P103/1.

Kohen E, Kohen C, Hirschberg JG, Wouters AW, Thorell B (1982b). The differential effects of the carcinogen dimethylnitrosamine in isocitrate and 6-phosphogluconate metabolism in single intact cells. Biochim Biophys Acta 720:424.

Kohen E, Kohen C, Rabinovitch A (1983a). Cell-to-cell communication in pancreatic islet monolayer cultures is modulated by agents affecting islet cell secretory activity. Diabetes 32:95.

Kohen E, Kohen C, Hirschberg JG, Wouters A, Thorell B, Westerhoff HV, Charyulu KN (1983b). Metabolic control and compartmentation. Cell Biochemistry and Function 1:3.

Kohen E, Kohen C, Hirschberg JG (1983c). Microspectrofluorometry of carcinogens in living cells. Histochemistry, in press.

Kopac MJ (1964). Micromanipulators. Principles of design operation and application. In Nastuk W L (ed) "Physical Techniques in Biological Research Vol V, Electrophysiological Methods, Part A," New York, Academic Press, p 191.

Krebs, HA (1973). Pyridine nucleotides and rate control. Symp. Soc. Exp. Biol. 27:299.
Loewenstein, WR, Kanno Y (1964). Studies on epithelial (gland) cell function, I. Modification of surface membrane permeability. J. Cell. Biol. 22:565.
Magee PN, Barnes JM (1967). Carcinogenic nitroso compounds. Adv Cancer Res 10:163.
Malaisse WJ, Malaisse Lagae F, Sener A (1982). The glycolytic cascade in pancreatic islets. Diabetologia 23:1.
Mansell JA, Clegg J (1983). Cellular and molecular consequences of reduced cell water content. Cryobiology, 20:591.
Mason JB (1959). Change transfer complexes in biological systems. Disc Faraday Soc 27:129.
McAllister RM, Meluyk J, Finklestein JZ, Adam EC, Gardner MB (1969). Cultivation in vitro of cells derived from a human rhabdomyosarcoma. Cancer 24:520.
Meda P, Kohen E, Kohen C, Rabinovitch A, Orci L (1982). Direct communication of homologous and heterologous endocrine in islet cells in culture. J Cell Biol. 92:221.
Michaels RL, Sheridan JD (1981). Islet of Langerhans': dye coupling among immunocytochemically distinct cell types. Science 214:801.
Moreno G, Salet C, Kohen C, Kohen E (1981). Penetration and localization of furocoumarins in single living cells studied by microspectrofluorometry. Biochem Biophys Acta 721:108
Moses V (1978). Compartmentation of glycolysis in Escherichia coli. In Srere PA, Estabrook RV (eds) "Microenvironment and Metabolic Compartmentation," New York: Academic Press, p 169.
Nastuk WL (1953). Membrane potential changes at a single muscle end plate produced by transitory application of acetylcholine with an electrically controlled microjet. Fed Proc 12:102.
Olson RA (1960). Rapid scanning microspectrofluorometer. Rev Sci Inst 31:844.
Racker E (1977). "A New Look at Mechanisms in Bioenergetics," New York: Academic Press, p 163.
Rousseau M (1957). Spectrophotometrie de fluorescence en microscopie. Bull Micr Appl 7:92.
Salmon JM, Thierry C, Serrou B, Viallet P (1978). Cinetique de decroissance et de transformation apparente du spectre de fluorescence de benzo(a)pyrene absorbe par

des cellules vivantes. Cas de lymphocytes et de macrophages. C R Acad Sci Ser D 287:345.

Salmon JM, Kohen E, Viallet P, Hirschberg JG, Wouters A.W., Kohen C, Thorell B (1982). Microspectrofluorometric approach to the study of free/bound NAD(P)H ratio as metabolic indicator in various cell types. Photochem Photobiol 36:585.

Sanford KK, Earle WR, Likely GD (1948). The growth in vitro of single isolated tissue cells. J Nat Cancer Inst 9:229.

Schachtschabel DO, Foley, GE (1972). Serial cultivation of Ehrlich ascites tumor cells in hypertonic media. Exp Cell Res 70:317.

Scott BR, Pathak MA, Mohn GR (1976) Molecular and genetic basis of furocoumarin reactions. Mutation Research 39:29.

Srere P, Estabrook R (1978). "Microenvironment and Metabolic Compartmentation," New York: Academic Press.

Stewart, WW (1978). Functional connections between cells as revealed by dye-coupling with a highly fluorescent napthalimide tracer. Cell 14:741.

Thorell B, Kohen E, Kohen C (1978). Metabolic rates and intercellular transfer of molecules in cultures of human glia and glioma cells. Med Biol Helsinki 56:386.

Veech R L, Eggleston LV, Krebs HA (1969). The redox state of free nicotinamide-adenine dinucleotide phosphate in the cytoplasm of rat liver. Biochem J 115:609.

Weber G (1961). Enumeration of components in complex systems by fluorescence spectrophotometry. Nature 190:27.

Welch GR (1977). On the role of organized multienzyme systems in cellular metabolism; a general synthesis. Prog Biophys Molec Biol 32:103

West SS (1965). Fluorescence microspectroscopy of mouse leucocytes supravitally stained with acridine orange. Acta Histochem Suppl 6:135.

Westerhoff HV, Hellingwerf KJ, Arents JC, Scholte BJ, Van Dam K (1981). Mosaic nonequilibrium thermodynamics describes biological energy transduction. Proc Natl Acad Sci USA 78:3554.

Westerhoff HV (1983). "Mosaic non equilibrium thermodynamics and the control of biological free energy transduction." Waarland, Netherlands: Drakkerij Gerja.

Whitlock JB, Miller H, Gelboin HV (1974). Aryl hydrocarbon (benzo[a]pyrene) hydroxylase induction in rat liver cells in culture. J Biol Chem 249:2616.

STUDIES ON THE ROLE OF Ca^{++} IN CELL DIVISION WITH THE USE OF FLUORESCENT PROBES AND QUANTITATIVE VIDEO INTENSIFICATION MICROSCOPY

Jesse E. Sisken[1], Robert B. Silver[2], George H. Barrows[3] and Sally D. Grasch[1]

Department of Pathology, College of Medicine, University of Kentucky, Lexington, Ky. 40536 [1]
Department of Biochemistry, Chicago Medical School, North Chicago, Ill. 60064 [2]
Department of Pathology, College of Medicine, University of Louisville, Louisville, Ky. 40232 [3]

I. Introduction

Much attention has been given to the possibility that calcium ions (Ca^{++}) play important roles in the regulation of cell division. This notion has been fostered by a variety of observations which include the Ca^{++}-sensitivity of microtubules in living cells (e.g., Marcum et al. 1978; Kiehart, 1981; Keith et al. 1983; Schliwa et al. 1981) and in isolated mitotic spindles (Salmon and Segall, 1980) as well as those assembled in vitro (Weisenberg, 1972). Ca^{++} has also been implicated in the formation and function of the actomyosin-containing contractile ring which is involved in cytokinesis in animal cells (for review, see Sisken, 1980). If it is true that calcium ions do play a regulatory role in mitosis, then systems must exist which regulate the temporal and spatial availability of these and, perhaps, other ions during this process.

An enormous literature has shown that there are a number of Ca^{++} regulatory systems present in eukaryotic cells. These include either membrane-bound organelles such as mitochondria, ER derivatives, and plasma membrane (for review, see Rasmussen and Waisman, 1981) as well as calcium binding proteins such as the ubiquitous calmodulin (for reviews, see e.g., Wallace et al. 1981; Means and Chafouleas, 1981). The location of membrane bound organelles in or near mitotic spindles or asters in a variety of cell types (e.g.Harris, 1962, 1978; Hepler, 1977, 1980; Silver et al. 1980; Wolniak et al. 1980, 1983;) provides a physical basis for models which relate the regulation of Ca^{++} to mechanisms by which Ca^{++} may in turn regulate the division process.

In order to establish that alterations in Ca^{++} levels are, in fact, controlling elements, it is important to determine if transient changes in levels of cytosolic Ca^{++} actually occur during the course of cell division. If such changes do occur, the question becomes one of how such transients are regulated. We are currently attempting to answer these questions in normal and transformed mammalian cells in culture and in cleaving sea urchin eggs. While the mitotic apparatus of mammalian cells and sea urchin eggs have very different gross morphologies, knowledge gleaned from both will yield important insights into the underlying mechanisms which regulate division in eukaryotic cells.

We are studying the alterations which occur in amounts and localization of membrane-associated Ca^{++} and calcium transport enzymes during the course of cell division. To these ends, we are using several complementary approaches which include morphological, quantitative cytochemical, biochemical and immunological analyses. One approach involves the use of chlorotetracycline (CTC) as a fluorescent probe for membrane-associated Ca^{++} in living cells. This agent has been so utilized in many biological systems such as pancreas (e.g., Taljedal, 1974; Chandler and Williams, 1978), leukocytes (e.g., Smolen et al. 1982) and a number of plant cells (e.g., Wolniak et al. 1980; Saunders and Hepler, 1981;) as well as in studies of calcium metabolism in isolated organelles (e.g., Caswell, 1972; Luthra and Olsen, 1976; Carvalho, 1978). The available data indicate that so long as it is used with some caution it is a valid and useful indicator of Ca^{++} transients (e.g., Binet and Volfin, 1975; Caswell, 1979; Millman et al. 1980; Caswell and Brandt, 1981).

A second probe we are using is an affinity purified polyclonal antibody directed against the calcium transport enzyme of guinea pig ileum smooth muscle. The location of calcium transport enzyme epitopes recognized by these antisera is determined by indirect immunofluorescence using a fluorescently labeled second antibody (Silver, 1983a, 1983b, 1984; Silver and Sisken, 1983).

For the detection and quantitation of fluorescence at the cellular level, we have been using a digital video image processing system coupled to a fluorescence microscope and a low light level video camera. We refer to this as a quantitative video intensification microscopy (QVIM) system after Willingham and Pastan (1978) who proposed the term video intensification microscopy (VIM) to describe a system in which they used a similar camera for qualitative studies. In this chapter we will briefly describe our QVIM system and some preliminary results we have recently obtained in studies with CTC and immunofluorescent probes in living and fixed cells, and in the isolated mitotic apparatus.

II. The QVIM System and Its Capabilities

For the work to be described, we have been using a Leitz Dialux EB 20 microscope fitted with an epifluorescence illumination system (Ploemopak 2.4) and a 50 watt mercury lamp. In the studies with CTC we used the Leitz D filter cube and, in addition, placed a 400 nm narrow band pass filter (Melles Griot, Irvine, Calif.) in the excitation light path. Direct photomicrography was done with a Leitz 35 mm camera. For QVIM work, images were obtained with a silicon intensifier target (SIT) video camera (Dage-MTI 65 MK II, Michigan City, Indiana). It has manual high voltage and gain controls and was fitted with adapters for a digital voltmeter (P. Mayernik, Leitz Inc. Cincinnati, Ohio) to permit reproducible setting of these parameters.

The image analysis unit has been described in detail (Barrows, et al, 1984). Briefly, it includes a gating system for video output which provides upper and lower threshold settings for noise limitation. In our latest version, this is done with calibrated potentiometers. It also includes a frame grabber with coupled video memory (Digital Graphics Systems, Palo Alto, Calif.). These are controlled by an S-100 bus-coupled Z-80 microprocessor system (Horizon-D, North Star, San Leandro, Calif.). The computer program was constructed in assembler blocks (Program Development System, (c) Allen Ashley, Pasadena, Calif.) which were called from a Basic language program for data retrieval and manipulation. These were used in conjunction with a light pen entry system

(Information Control Corp. Los Angeles, Calif.), a CRT console (Soroc, Anaheim, Calif.) and an 80-column printer (MPI, Salt Lake City, Utah). Images were displayed in black and white or in pseudocolor on a Zenith NTSC color monitor.

In practice, we call up a live image on the monitor and obtain a digitized image which is held in computer memory and displayed immediately on the monitor. Different intensities can be displayed in a number of ways including in shades of gray, shades of a single color or, for superior visual recognition of intensity differences, with the assignment of contrasting colors to different intensities. For localization studies, these images can be recorded by photographing the monitor screen using a tripod-mounted 35mm camera. For quantitative work, the light pen is used to delimit areas from which measurements are desired. The picture elements (pixels) within the delimited area are counted and their intensities determined. These are then integrated to produce a value for total intensity which we term video integral. The output obtained also includes the area and perimeter of the selected area, the number of pixels with light intensities above background and the distribution of pixels as a function of light intensity over a 16-level gray scale.

The system is menu driven and uses a minimum of commands or keystrokes for initial boot-up, image grab and image recording. The use of the light pen further simplifies the interaction between the microscopist and the QVIM system. Finally, data may be obtained "on-line" from live images or from images recorded on magnetic media, i.e., floppy discs or video tape.

A QVIM system should meet several criteria to establish its efficacy for the quantitation of fluorescence images. These include a linear response to fluorescent light output, a high degree of reproducibility and ease of operation when used with live or recorded images. In this section, we will briefly describe the tests we have done to demonstrate that our system meets these criteria. Details of these tests are described elsewhere (Barrows, et al.1984).

The linearity of response was determined by measuring the SIT camera output from serial dilutions of fluorescein solutions which were placed under the coverslip of a hemocytometer which provides a fixed sample path length. Using the tungsten lamp of the microscope, we focussed the microscope halfway between the upper surface of the slide and the bottom surface of the coverslip. We then switched from visible, transmitted light to the mercury epifluorescence light source and measured fluorescence light output in a fixed area of the field. In order to keep the light output from the higher fluorescein concentrations within the dynamic range of the system, neutral density filters were used in the exciter light pathway and the data outputs were corrected for the known optical densities of the filters. The data showed that under these conditions, the system has excellent linearity over the fluorescein concentration range of 2×10^{-8} to 1×10^{-3} M.

The reproducibility of the system was tested by measuring the fluorescent light emitted by two different standard samples: glutaraldehyde-fixed chick red blood cells and fluorescent beads (Fluorospheres, Coulter Electronics, Elks Grove Village, Ill.). Unimodal peaks were obtained from both samples with coefficients of variation [CV's (standard deviations as percentages of the mean)] similar to those obtained by flow microfluorometric techniques for these same kinds of samples. For fluorescent beads we obtained CV's between 3 and 4% which is within range of the 2% reported for these materials by other workers (e.g., McCutcheon and Miller, 1979). For the chick RBC's, we obtained CV's of 18.7 % which are comparable to the 20% range reported elsewhere (Becton-Dickinson Instruction Manual for FACS3 Cell Sorter/Analyzer, Mountain View, Calif.)

To demonstrate that the system is applicable to biological materials, we measured the light output from human diploid fibroblasts stained with the DNA-specific Hoechst 33258. Histograms showing the amount of fluorescence per cell had the expected configurations. The distributions were bimodal for both cell strains with major peaks corresponding to the pre DNA-synthetic, 2C amount of DNA and minor peaks at twice this value corresponding to the post DNA-synthetic 4C value.

Thus, the results of these tests indicate that our QVIM system has good linearity, can yield reproducible measurements and can provide biologically meaningful data.

III. Preliminary Findings

A. Studies With Chlorotetracycline in HeLa Cells

As described earlier, CTC has been used as a probe for membrane-associated Ca^{++} in numerous studies) but in most cases little was done to determine optimum conditions for staining intact cells or to examine the biological effects of the probe. Simple microscopic observation indicated to us that the intensity of fluorescence was both concentration and time dependent. QVIM measurements of cells exposed to various concentrations of CTC for 1 hour are shown in fig. 1. The increase in intensity is clear but it is not linear over the range of 10-100 uM. The time dependency of staining is demonstrated in fig. 2. Here, emission intensity is presented as a function of time for cells stained with 10 uM CTC. It can be seen that staining intensity increases rapidly for about 1-1.5 hours and then declines even though

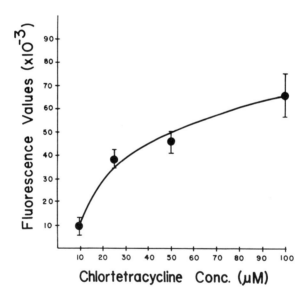

Fig. 1. Fluorescence intensity as a function of CTC concentration. HeLa cells growing in Rose chambers were exposed to various concentrations of CTC in Eagle's Minimal Essential Medium (MEM) with 2% calf serum for one hour. Fluorescence intensities for individual interphase cells were determined by QVIM.

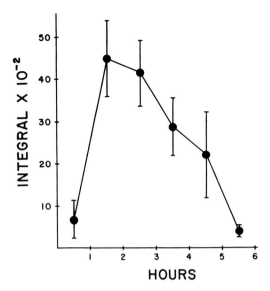

Fig. 2. Time dependency of CTC staining. HeLa cells growing in Rose chambers were exposed to 10 uM CTC in Eagle's MEM with 2% calf serum. At the indicated times afterward, fluorescence intensities were determined for individual metaphase cells by QVIM.

cells had been kept in the dark. Other unpublished data indicate that the rate of this decline is also a function of initial CTC concentration and may result from the metabolic breakdown of CTC itself.

In our initial studies, we used CTC concentrations ranging from 50-100 uM as was done by previous workers (e.g., Chandler and Willams, 1978; Schatten et al. 1982) and our observations were consistent with their reports. Specifically, we saw the bright punctate fluorescence attributed to mitochondria and a relatively bright, diffuse fluorescence thought to emanate from the endoplasmic reticulum. We also observed brightly fluorescent nuclear envelopes. However, the punctate appearance of mitochondria suggested that the cells were experiencing CTC-induced damage (Sisken and Grasch, 1983; Sisken, et al, 1983) since under normal conditions, mitochondria in these cells should be filamentous rather than round and punctate.

Prompted by the apparently artifactual changes in cell structure following the use of CTC, we undertook a detailed study of CTC's biological effects (Sisken and Grasch, 1983; Sisken et al. 1983). Our findings are summarized by the following 5 points: (1). Levels of CTC above 25 uM can inhibit the growth rate of HeLa cells. (2). Mitochondria were damaged by CTC and, indeed, levels as low as 10 uM could cause ultrastructural changes if cells were exposed to the agent for as long as 2 hours. (3). The nuclear envelope-endoplasmic reticulum complex can also be altered as a function of CTC concentration and duration of treatment. (4). These alterations in the NE-ER complex account for the bright, diffuse, cytoplasmic fluorescence seen at higher CTC levels. (5). If exposures were restricted to concentrations of 10 uM for one hour, the mitochondria were the main sources of fluorescence and they retained their normal filamentous shapes (Sisken and Grasch, 1983; Sisken, et al, 1983). From these studies, we concur with the suggestion of Sehlin and Taljedal (1979) that when using CTC as a probe for membrane-associated Ca^{++}, exposures should be kept as low as possible, i.e., concentrations should be no more than 10 uM and exposure times should be as short as possible.

Examples of digitized images of interphase and mitotic HeLa cells are shown in fig. 3. When HeLa cells at metaphase are exposed to either high or low levels of CTC, the mitotic spindle remains nonfluorescent while extraspindle zones are brightly fluorescent (fig. 3c). This indicates that mitotic spindles in these cells contain little detectable membrane-associated Ca^{++}. This is consistent with published electron micrographs which show that these spindles contain relatively few membranous elements (e.g., Robbins and Gonatas, 1964). As division progresses, the fluorescent organelles become redistributed so that by late telophase, most of the cell's fluorescence is concentrated in the region of the cytokinetic furrow (fig. 3e and Sisken et al. 1981). Such observations are consistent with models of mitosis which suggest: (1) that free Ca^{++} must be removed from the vicinity of the spindle early in the process to facilitate microtubule assembly and (2) that at about late anaphase, Ca^{++} must be released in the interzonal regions to bring about breakdown of the remnants of the mitotic spindle and to stimulate the actomyosin-containing microfilaments of the contractile ring (for review, see Sisken, 1980).

If such alterations in Ca^{++} levels are a result of the

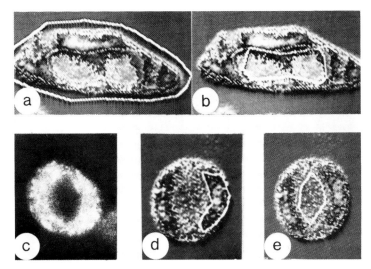

Fig. 3. Fluorescence images of HeLa cells exposed to CTC. Figs. 3a and 3b are of the same digitized image of an interphase cell. Fig. 3a shows the entire cell outlined with the light pen while in fig. 3b only a portion of the nucleus was delimited. Fig. 3c shows a live metaphase cell exposed to a high concentration of CTC (100 uM). The mitotic spindle is oriented horizontally. Figs. 3d and 3e are of a digitized image of a different metaphase cell oriented the same way showing the use of the light pen for the selection of small portions of the cell.

sequestration and release of the ion from intracellular sources, then these models predict that quantitative changes in amounts of membrane-associated Ca^{++} should occur as cells progress through mitosis. Wolniak et al (1980), observed CTC fluorescence between spindle fibers in metaphase Haemanthus endosperm cells and have more recently reported a transient decrease in CTC fluorescence in the polar zones of Haemanthus spindles just prior to the onset of chromosome movement (1983).

We have now begun two types of studies to determine if transient changes in membrane-associated Ca^{++} occur in HeLa cells. First we are measuring fluorescence in populations of cells at each stage of division and determining whether average fluorescence values vary from stage to stage. Secondly, we are making repeated measurements on individual cells as they progress through division. The results of such experiments, when completed, should help us understand what is happening to membrane-associated Ca^{++} pools during the course of mitosis.

B. Calcium Sequestration in Sea Urchin Mitotic Apparatus

In contrast to the spindles of HeLa cells, the mitotic apparatus of cleaving sea urchin embryos contains large numbers of membranous organelles derived from the endoplasmic reticulum. These have been isolated and shown to be able to sequester Ca^{++} in an ATP-dependent fashion (Silver et al., 1980; Silver, 1983b). The Ca^{++}-sequestering capabilities of isolated, membrane-containing sea urchin mitotic apparatus is seen in table 1. These preparations are free of mitochondria and plasma membrane (Silver et al., 1983).

Table 1. Uptake of ^{45}Ca by isolated native mitotic apparatus.

Additive	Radioactivity incorporated into pelleted mitotic apparatus 10^4 cpm
None	21.19
ATP	38.92
ATP + A23187	18.38
AMPPCP	23.64
AMPPNP	20.03

See Silver et al. (1980) for details. With permission of the MIT Press.

The findings from sea urchin, Haemanthus and HeLa cells are all consistent in that in each case there appear to be Ca^{++}-sequestering organelles located within, or in the vicinity of, the mitotic spindle (as well as in asters in the case of the sea urchin eggs). In Haemanthus (Hepler, 1980; Wolniak et al. 1980), and sea urchins (Harris, 1962; Silver, 1980) as well as in spider spermatocytes (Wise and Wolniak, 1983), CTC-staining, membranous organelles are present within the spindle. On the other hand, HeLa cells are different in that such elements are comparatively rare in the spindle per se but are found immediately outside the spindle. This difference might be due to the fact that HeLa spindles are so much smaller which means that diffusional distances and possibly the amounts of Ca^{++} that have to be sequestered are much less. Thus, cells like HeLa may be able to regulate the intraspindle environment without the necessity of having sequestering elements within the spindle itself.

C. Studies with Antibodies Against Calcium Pump Enzymes

We have begun to use our QVIM system to study the localization of ATP-dependent calcium transport enzyme in dividing mammalian cells and in the isolated, native mitotic apparatus of sea urchins. This is being done in conjunction with our studies on the quantitation and localization of membrane-associated calcium. One of our goals is to determine the degree of coincidence of localization of membrane-associated calcium and the calcium transport enzyme in these cells.

Antibodies to the calcium transport enzyme of guinea pig ileum smooth muscle do cross react with the calcium sequestering vesicles of the mitotic apparatus of sea urchin embryos in vitro and in vivo (Silver, 1983a, 1983b; Silver, 1984). These affinity purified IgGs also inhibit calcium sequestration in vitro and cross react with antigens in both mitotic and interphase HeLa cells. Further, in support of the view that the calcium transport enzyme and a pool of membrane-associated Ca^{++} are parts of the same system, our preliminary results indicate that CTC-staining organelles tend to co-localize with these antibodies in both sea urchins and HeLa cells (Silver 1983a, 1984; Silver and Sisken, 1983).

IV. Summary

It is clear that QVIM systems, when combined with appropriate fluorescent probes can be utilized to perform quantitative cytochemical studies on living and fixed cells. They also have the potential to facilitate studies of substances which like Ca^{++} are not easily studied by other means. The preliminary studies we have described support the idea that calcium ions and calcium transport enzymes may indeed play important roles in cell division and indicate that the tools we have at hand should help us further our understanding of the mitotic process.

Acknowledgments

The work presented in this chapter was supported in part by grant CA27399 from the National Cancer Institute, National Institutes of Health (JES), grant CD-128 from the American Cancer Society (RBS) and grants from the American Cancer Society, Illinois Division (RBS), the Chicago Heart Association (RBS) and by BRSG grants to UHS/CMS and the University of Kentucky.

References

Barrows GH, Sisken JE, Allegra JC, Grasch SD (1984). Measurement of fluorescence using digital integration of video images. (submitted).
Binet A, Volfin, P (1975). Effect of the A23187 ionophore on mitochondrial membrane Mg^{2+} and Ca^{2+}. FEBS Lett 49:400.
Carvalho CAM (1978). Chlorotetracycline as an indicator of the interaction of calcium with brain membrane fractions. J Neurochem 30:1149.
Caswell AH (1972). The migration of divalent cations in mitochondria visualized by a fluorescent chelate probe. J Membrane Biol 7:345.
Caswell AH (1979). Methods of measuring intracellular calcium. Int Rev Cytol 56:145.
Caswell AH, Brandt NR (1981). Ion-induced release of calcium from isolated sarcoplasmic reticulum. J Membrane Biol 58:21.
Chandler DE, Williams JA (1978). Intracellular divalent cation release in pancreatic acinar cells during stimulous-secretion coupling I. Use of chlorotetracycline as fluorescent probe. J Cell Biol 76:371.
Fabiato AA, Fabiatto F (1979). Use of chlorotetracycline fluorescence to demonstrate Ca^{2+}-induced release of Ca^{2+} from the sarcoplasmic reticulum of skinned cardiac cells. Nature 281:146.
Hallett M, Schneider AS, Carbone E (1972). Tetracycline fluorescence as calcium-probe for nerve membrane with some model studies using erythrocyte ghosts. J Membrane Biol 10:31.

Harris P (1962). Some structural and functional aspects of the mitotic apparatus in sea urchin embryos. J Cell Biol 14:475.

Harris P (1978). Triggers, trigger waves and mitosis: a new model. In Buetow DE, Cameron IL, Padilla GM (eds): "Cell Cycle Regulation," New York: Academic Press, p. 75.

Hepler PK (1977). Membranes in the spindle apparatus: their possible role in the control of microtubule assembly. In Rost TL, Gifford EM Jr. (eds): "Mechanisms and Control of Cell Division," Stroudsburg, Pa. Dowden, Hutchinson and Ross, Inc. p.212.

Hepler PK (1980). Membranes in the mitotic apparatus. J Cell Biol 86:490.

Keith C, DiPaola M, Maxfield FR, Shelanski ML (1983). Microinjection of Ca^{2+}-calmodulin causes a localized depolymerization of microtubules. J Cell Biol 97

Kiehart DP (1981). Studies on the in vivo sensitivity of spindle microtubules to calcium ions and evidence for a vesicular calcium-sequestering system. J Cell Biol 88:604.

Luthra R, Olsen MS (1976). Studies of Mitochondrial calcium movements using chlorotetracycline. Biochim Biophys Acta 440:744.

Marcum JM, Dedman JR, Brinkley BR, Means AR (1978). Control of microtubule assembly-disassembly by calcium-dependent regulator protein. Proc Nat Acad Sci 75:3771.

Means AR, Chafouleas JG (1981). Regulation by and of calmodulin in mammalian cells. Cold Spring Harb Symp Quant Biol 46:903.

McCutcheon MJ, Miller RG (1979). Fluorescence intensity resolution in flow systems. J Histochem Cytochem 27:246.

Millman MS, Caswell AH, Haynes DH (1980). Kinetics of chlorotetracycline permeation in fragmented, ATPase-rich sarcoplasmic reticulum. Memb Biochem 3:291.

Rasmussen R, Waisman D (1981). The messenger function of calcium in endocrine systems. In Litwack G (ed): "Biochemical Actions of Hormones," vol. III. New York: Academic Press p.1

Robbins E, Gonatas NK (1964). The ultrastructure of mammalian cells during the mitotic cycle. J Cell Biol 21:429.

Salmon ED, Segall RR (1980). Calcium labile mitotic spindles isolated from sea urchin eggs (Lytechinus variegatus). J. Cell Biol. 86:355.

Saunders MJ, Hepler PK (1981). Localization of membrane-associated calcium following cytokinin treatment in Funaria using chlorotetracycline. Planta 152:272.

Schatten G, Schatten H, Simerly C (1982). Detection of

sequestered calcium during mitosis in mammalian cell cultures and in mitotic apparatus isolated from sea urchin zygotes. Cell Biol Internat Repts 6:717.

Sehlin J, Taljedal I (1979). $^{45}Ca^{2+}$ uptake by dispersed pancreatic islet cells: effects of D-glucose and the calcium probe, chlorotetracycline. Pflugers Archiv 381:281.

Silver RB, Cole RD, Cande WZ, (1980). Isolation of mitotic apparatus containing vesicles with calcium sequestration activity. Cell 19: 505.

Silver RB (1983a). Co-localization of calcium sequestered in vivo and the calcium transport enzyme in isolated sea urchin mitotic apparatus. J Cell Biol 97:41a.

Silver RB, (1983b). Inhibition of mitotic anaphase and cytokinesis and reduction of spindle birefringence following microinjection of anti-calcium transport enzyme IgGs into Echinaracnis parma blastomeres. Biol Bull 165:495.

Silver RB, (1984). Inhibition of mitosis, reduction of spindle birefringence and relaxation of the cleavage furrow by antibodies directed against the calcium transport enzyme. (submitted).

Silver RB, Sisken JE (1983). Coincident localization of membrane-associated calcium and the calcium transport system in mitotic and interphase HeLa cells. J Cell Biol 97:41a.

Sisken JE (1980). The significance and regulation of calcium during mitotic events. In Whitson G (ed): "Nuclear-Cytoplasmic Interactions in the Cell Cycle," New York: Academic Press, p. 271.

Sisken JE, Grasch SD (1983). Some effects of the Ca^{++} probe, chlorotetracycline, on HeLa cells. J Cell Biol 97:42a.

Sisken JE, Geissler R, Grasch SD (1983). The effects of chlorotetracycline on the ultrastructure of HeLa cells. J Cell Biol 97:161a.

Smolen JE, Eisenstat BA, Weissman G (1982). The fluorescence response of chlorotetracycline-loaded human neutrophils. Correlations with lysosomal enzyme release and evidence for a trigger pool of calcium. Biochim Biophys Acta 717:422.

Taljedal I (1974). Interaction of Na^+ and Mg^{2+} with Ca^{2+} in pancreatic islets as visualized by chlorotetracycline fluorescence. Biochim Biophys Acta 372:154.

Wallace RW, Tallant EA, Cheung WY (1981). Multifunctional role of calmodulin in biologic processes. Cold Spring Harb Symp Quant Biol 46:893.

Weisenberg RC (1972). Microtubule formation in solutions containing low calcium concentrations. Science 177:1104.

Willingham MC, Pastan I (1978) The visualization of fluorescent proteins in living cells by video intensification microscopy (VIM). Cell 13:501.

Wise DA, Wolniak SM (1983). The intraspindle membrane system of spider spermatocytes is rich in associated calcium. J Cell Biol 97:41a.

Wolniak SM,, Hepler PK, Jackson WT (1980). Detection of the membrane-calcium distribution during mitosis in Haemanthus endosperm with chlorotetracycline. J Cell Biol 87:23.

Wolniak SM, Hepler PK, Jackson WT (1983) Ionic changes in the mitotic apparatus at the metaphase/anaphase transition. J Cell Biol 96:598.

IMAGE ANALYSIS IN BIOMEDICAL RESEARCH

RONALD R. COWDEN
DEPARTMENT OF BIOPHYSICS
QUILLEN-DISHNER COLLEGE OF MEDICINE
EAST TENNESSEE STATE UNIVERSITY
JOHNSON CITY, TENNESSEE 37614, U.S.A.

INTRODUCTION

Image analysis as we know it today is an outgrowth of the space age high technology that married the television camera to the computer. While there are other ways to produce regularly spaced rectangular two dimensional digitized matrices, video cameras offer the most economical and convenient way to achieve this. Again in contemporary context, most of us think of image analysis as some combination of density or intensity information coupled with geometric data which can be combined into complex algorithms, but there are forms of image analysis that depend entirely on the manipulation of geometric data, and which are suited to both digitized tablets and video systems. Morphometry was practiced in various ways first with the light, then later the electron microscope, using micrometer oculars or planimeters on projected images or photographs to determine areas (see Aherne and Dunnill, 1982). While this approach has been useful in some cell biological or macromorphometry applications, it has been particularly useful in neuroanatomy where the cytoarchitecture of nests of neurons, the "nuclei" of the CNS, can be characterized by ranges of perikaryon areas or nuclear/perikaryon ratios. For this reason the use of digitized tablets interfaced to dedicated microcomputers has become common in neurobiology groups (see Cowan and Wann, 1973). As an extension, programs now allow three-dimensional reconstruction of nuclear cytoarchitecture from tracings of serial sections or the mapping of axonal arborizations.

In the same sense stereology was devised, principally by Weibel (1979, 1980) first as a method for analysis of cells and tissues at the light microscope level, then it was subsequently extended to analysis of cell organelles at the ultrastructure level. The approach depends on the number of encounters between a regularly spaced grid of lines and targeted structures in the field, and allows the calculation of the partial specific area occupied by that target with a high degree of confidence. When stereology was transferred to ultrastructure it became extremely useful in characterizing morphometry in general. Stereology can be performed using an ocular containing appropriate reticules, with a digitized tablet or with a television system interfaced to a computer. However, as a dependence on the more complex manipulation of image information unfolded, special new forms of logic were required. The Fontainebleau School in France developed a logic system that was covered in a treatise by Serra (1982), and which is currently used in a number of commercial systems. Other developers or investigators have developed their own special image analysis logic. The recent publication of Lockart et al., (1984) offers a good example of the mixture of density and geometric data this approach implies. The case in point was a study of Feulgen stained nuclei of rabbit lymphocytes, macrophages and stimulated macrophages; with an attempt to discriminate between these nuclei on the basis of nuclear textures. The first three factors: integrated absorbance, absorbing area and the mean of these absorbing values fall within the perview of conventional quantitative cytochemistry. However, the integration of pixels with optical densities (O.D.) greater than 1.4, the determination of the ratio of the integrated pixels with O.D. > 1.4 divided by nucler area (number of pixels above background), or the determination of the nuclear center and the density of pixels in concentric radii and their further manipulation constitutes a good and readily understandable example of the type of logic used in computer analysis of video images. Six parameters were identified that resulted in appropriate assignment of nuclear classes, and the use of three sets of combinations of three parameters allowed assignments of nuclei to the morphological class of cell at the level of better than 80 and up to better than 90 percent certainty. In addition they pointed out that "heterochromatin" should be redefined as an ultrastructural feature relating to the 30 n.m. diameter chromatin fibers, and that chromatin condensation viewed at the light microscopic level is probably a consequence of differences in nuclear matrix organization rather than differences in transcriptional activity per se. There are parameters in use in image analysis that represent characteristics extracted by the exercise of logic and experimental experience. In considerations limited to geometric terms only these might include determination of the longest cord, center of gravity, convex perimeter, perimeter divided by radius squared, etc. With the added consideration of optical density or intensity, the permutations increase enormously, and shortly achieve levels of complexity that could only be correlated within a set of measurements of populations of cells with the aid of a computer.

Image analysis approaches to the description of nuclear texture are not limited to video camera systems. Mello (1983) performed several image analysis operations concerned with chromatin texture using a Zeiss (Oberkochen, FRG) microspectrophotometer equipped with a stepping stage, while Boss and Mays (1976) and Boss et al.(1976) used similar approaches to evaluate Feulgen stained nuclei with an image scanning Vickers M-85 (Vickers Instruments, York, UK). In both instances, the systems employed were operating with fewer picture points than would have been available from a video image, but the sampling was sufficient to allow the achievement of some reasonably secure conclusions concerning chromatin texture.

The main impetus for the commercial development of image analysis systems, most of them using television cameras to generate the image, emerged from two clinical applications: the desirability of developing marketable instruments that would efficiently discriminate between white blood cell types (human blood leucocyte differential counts), and instruments that might be used to automate cytodiagnosis of cancer--particularly on exfoliative preparations of vaginal cervix smears. Since there are several blood cell differential count systems in the field that have even penetrated to the level of community hospital pathology laboratories, it seems that this approach has been successful. In most cases 200 randomly encountered leucocytes are characterized. The ones that do not fit the classification algorithms are marked, returned to by an X-Y stepping stage from coordinates mapped in memory and assigned a definitive category by the technician by visual inspection. The classification of each type of cell is achieved by multiparameter comparisons that include at least: integrated nuclear absorbance, the ratio of nuclear to cytoplasmic area, and cytoplasmic absorbance at two wavelengths. Some "form factors" are also included. At the least, the system is required to take and store images at two wavelengths, extract the appropriate parameters and compare these to boundry values for each definitive type of cell in a "look-up table" in the computer software.

In the case of exfoliative cytology, the problem becomes more complicated. In some types of cancer, most of the cells contain unbalanced chromosome sets. If DNA per nucleus is measured, in many types of cancer the malignant nuclei produce values that fall between G_1 (2C=Diploid) and G_2 (4C=tetraploid) for the species, or above G_2. There are some types of malignancies in which DNA values do not change remarkably, and these must be approached with other markers. However, the determintion of premalignant conditions and the evaluation of these cells for malignant potential is complicated by the fact that dysplasias and hyperplasias are characterized by diploid G_1 DNA values; but there are changes in chromatin texture, nuclear size, nuclear/cytoplasmic ratio and the number and disposition of nucleoli. The same is true for certain virus induced tumors. Thus in cancer diagnosis, it has become important to sort out a range of factors other than DNA amounts that might allow confident assignment of a given cell to one category or another (see Wied et al., 1982). There are at least two major and divergent philosophical approaches to this problem: (1) An all-out multiparameter high-resolution analysis of conventionally prepared smears in which complex algorithms for texture are applied, and which would tend to be preferred in the sense that it would not alter the current clinical conventions of preparing and staining smears. (2) The development of special cytochemical methods that would evaluate DNA per nucleus and some combination of other markers that might be approached by microspectroscopy. In general, the former requires mainframe computing capacities, 512 X 512 X 8 to 12 bit images with very large frame memories, while the second approach can be considered using simpler systems.

DETECTORS AND SCANNERS

The principal detector used in most image analysis systems today is the television camera because it inherently generates a two dimensional raster matrix of picture points, and their use is reasonably convenient and economical as compared to some other input options. Some very early cytochemical applications of television systems are given in Williams (1964) and Montognery et al., (1964). However, it should be noted at the outset that television cameras are not absolutely ideal light intensity detectors because they display some deviations from linearity, and because the video pixel architecture enforces some distortions from ideal geometric representation. These problems can be compensated in software. But video cameras may also produce errors because the charge is not completely drained off the imaging phosphor after each sweep of the electron beam (static charge wiping errors), or a very bright spot may overflow a line and affect adjacent lines ("blooming"). Nevertheless, in comparison with other options, video cameras are the most convenient and economical image scanning devices available at this time, and the problems they present can be sufficiently overcome to allow their use in a broad range of applications.

In most brightfield applications, Newbicon or other types of video tube cameras are selected because they are more sensitive than Vidicon tubes, and offer better signal-to-noise characteristics. As in photomultiplier based absorbance measurements, an enhancement in accuracy can be achieved if the image is averaged. In some cases, the image is scanned 256 times or more, and the average value for each pixel is recorded. This requires a video tube that will rescan the same pattern with some consistancy (low "jitter"). For most fluorescence applications it is necessary to use an intensified camera, or a silicon intensified target (S.I.T.) Videcon camera, which typically produce gains in sensitivity over a standard Vidicon camera of 750 to 1500 times. If U.S. standard video (EIA-RS-170) is used, the images are presented at the rate of 60 frames a second, but if an interlace scan is inserted between the original scanning pattern to increase resolution by reducing the area of each picture point to one-fourth that of a standard pixel, the rate drops to 30 frames a second. It is possible to use 3X interlace to generate four interlaced raster scans which further reduce the pixel areas, but this also reduces the number of frames per second to a level that causes a distinct flicker. The highest resolution scanning should only be used in the "stop frame" mode in which the frames are presented consecutively, one at a time, not continuously as in real-time video. These various modes; i.e., standard, single interlace and 3X interlace; most nearly correspond to 256 X 256, 512 X 512 and 1024 X 1024 pixels, and are accordingly stored in banks of high-speed memory of corresponding sizes. It must be obvious that very large memories are required to store these frames; and that with each upward step in resolution, far larger memories are required. Beyond this, the density or intensity data must be managed. In some useful systems, only 4 bits of digitization are used (16 grey levels), while others extend upward to 12 bits (4096 grey levels) and some 16 bit systems are in use in experimental systems. In most biomedical applications, image analysis systems have been designed for 512 X 512 pixel resolution and up to 8 bit digitization. This requires a relatively large amount of high-speed memory, particularly if background is subtracted, "windows" are generated and some image editing capacities are included.

Television cameras are not the only input devices that rapidly generate two dimensional matrices of digitizable picture points. In due course two dimensional diode arrays might well fulfill this requirement and do so more effectively than television cameras. Those currently available, even the intensified diode arrays, lack sufficient sensitivity or resolution to be used in many applications- particularly in fluorescence, but they can be used in some applications that will tolerate a 128 X 128 or 256 X 256 pixel resolution and lower inherent sensitivities. A small array has even been used in a fluorescence application to determine "flashes" in cell bodies of neurons fluorochromed with dyes sensitive to changes in membrane electrical potential (Grinvald et al., 1984). Diode arrays can avoid the problems of "static wiping" and "bloom" encountered in video cameras, and can be scanned at standard video rates. The results can be displayed directly on a standard or high-resolution television monitor. Very high resolution scanning patterns can be obtained by using a linear array of diodes to provide one axis, and stepping either the image or the array in the other. However, this method is relatively slow, particularly if pixel values are to be averaged. While this works extremely well in the absorbance mode, it is too slow for most fluorescence applications. Since not all the diodes within an array output the same value with a given amount of light falling upon it; each array must be evaluated and "normalized" to a baseline value. While diode arrays represent the future in image analysis, their time has not quite come for most contemporary image analysis applications.

There are scanning devices that depend on rotating mirrors or prisms, and probably the fastest and most accurate raster generating systems depend on the use of hexagonal nut spinning mirrors (see Bartels et al, 1982). These tend to be relatively expensive and temperamental, but they can operate at sampling rates greater than video speed without the dimensional limitations of video formats and have been incorporated into some main-frame based applications. The older conventions of stepping the object in 0.25 to 0.5 um steps with an X-Y stepping stage has already been noted, and scanning of a projected image of an object with a set of rotating prisms is used in the Vickers M85 scanning microdensitometer. It is also possible to scan a point source of well collimated light in a raster pattern, and this approach was used by Myall and Mendelsohn (1970) in the development of the CYDAC system used at Lawrence Livermore Laboratories in Livermore California. This system employs a video raster of sufficient brightness and spectral composition to generate a flying spot that is focused on the object with the condenser lens. All of these systems depend on photomultiplier tube based photometers to measure light. Photomultiplier tube measurements are probably more accurate than video measurements and are certainly more sensitive, but most photomultiplier tubes are not sensitive enough to allow measurement of fluorescence emissions from 0.5-1.0 um^2 areas at the specimen level even when arc discharge lamp excitation sources are employed. In absorbance measurements, the light level can be altered and signal averaging can be employed, but the only recourse open in fluorescence is to move to more powerful laser sources with greater luminous fluxes of light. This has been tried experimentally, but not thus far in any commercial designs. While photomultiplier tube based systems work very well for measurement of fluorescence emissions from whole nuclei or cells, they leave a good bit to be desired in scanning fluorometry applications.

THE COMPUTERS

The heart of image analysis systems lie in the design of the computer, its associated peripherals and software. In very large systems, large banks of high-speed memory are used in conjunction with array processors to "segment" the images; i.e., to reduce the images to useful numerical parameters. Most of the software written for the mainframe or large minicomputer image analysis systems are in FORTRAN IV, and those who work with these systems find it desirable to have at least one member of the research group with expertise in image analysis and FORTRAN IV programming. With the development of some larger microprocessors, the introduction of "pipeline" frame memories and less expensive array processors, there has been a tendency to program in PASCAL, a more structured higher-level language for larger mini and some microcomputers. However, some very fine systems have been developed using 8-bit microprocessors extended with frame memory (high-speed memory that will store digitized frames) and graphics controller boards. Two small systems have recently been described in the literature. Both use small microprocessors to evaluate fluorescence detected with a SIT-Videcon camera, but the fundamental approaches are different. Tanasugarn et al. (1984) used a Hamamatsu (Hamamatsu Systems, Inc. Waltham, MA 02254)Polyprocessor which is essentially a specialized computer designed to execute many hard-wired functions relating to the manipulation of Hamamatsu TV camera generated image data. A smaller microprocessor is used to handle some switching and statistical analysis functions. Barrow et al., (1984) have used a Northstar microcomputer based on the Z-80 chip with an add-on frame memory and light pen and a Dage-MTI SIT Videcon TV camera. Most of the small systems use a combination of BASIC and ASSEMBLER language in which the general program is formatted in BASIC, but ASSEMBLER subroutines are used where speed is important. As useful as the small systems might be, they do not have the capacity to cope with 24 frames of the same nucleus at different evenly spaced focal levels, smoothing of these images (removal of glare), and connection of the most probable ends between images and drawing of the whole nucleus containing four polytene chromosomes in three dimensions

(Mathog et al., 1984). This required a VAX at the lower edge of mainframe systems, a large image manipulation library and an array processor. Similarly, data reduction of the kind attempted by Preston's group (see Tourassis et al., 1983, Wied et al., 1982) requires a large computer capacity. While it seems certain that most biologists will more probably have access to one of the smaller systems, the limitations of these systems should be held in perspective.

PREPARATIVE AND STAINING REQUIREMENTS

This is almost a case of first things last. A system is only as good as the material it evaluates. With enhanced sensitivity, it has become essential that more attention be given to the preparation and staining of cells and tissues. In the case of blood smears, various devices have been developed which allow preparations of smears of uniformly spread cells (see Bacus, 1974). While the method of centrifuging a diluted drop of heparinized or EDTA treated blood works reasonably well for mammalian material, the viscosities of blood in non-mammalian vertebrates differ, and this approach has not been satisfactory in those applications. Various forms of cytocentrifugation have been attempted, and the method of Lief et al. (1971) has proven to be one of the more useful methods of preparing cells or nuclei from suspensions. It was particularly noted that when preparations remained "wet", staining patterns were far more consistent than in cases where the cells or isolated nuclei were allowed to dry (Curtis and Cowden, 1980). Another approach has been the collection of cells on a sub-micron membrane filter which is subsequently stained or from which cells are transferred to a cleaned slide with reverse pressure (Oud et al., 1984).

The composition and conditions of staining provide yet another problem in high resolution image analysis. The thiazan-eosin Romanovsky stains conventionally used in hematology have been a continuing source of concern, particularly since both classes of dyes are frequently contaminated in commercial samples. A great deal of work has been done on the purification, spectral properties and staining conditions for these complex mixtures (see Marshall and Galbraith, 1984; Ruter et al., 1982). Similarly, the composition of the Papanicoloau stain, used in exfoliative cytology, has also been a subject of the same kind of study to improve the reproducibility of clinical specimens used in image analysis (Galbraith et al., 1979), The alternative has been to look for combinations of histochemical stains that would result in the selective demonstration of one class of cell as compared to another (see Cornelisse and Ploem, 1976), but at this point, this approach has been limited to research applications.

CONCLUDING REMARKS

With the marriage of video images and the computer, image analysis has arrived in the bio-medical sciences. In practice, its application has been heavily medical with the principal emphasis on blood cell differentiation or diagnosis of cancer cells and to a lesser extent for karyotype analysis (see Pieper, 1982) or studies in cell biology. Its extension into general biology per se has been minimal, but the options this technology offers are enormous. The objective of this section has been to introduce image analysis, and some of the concepts used in image analysis, to biologists in the hope that it would remove some of the mystique associated with this technology; and to present this a level that would offer an entry into image analysis literature and allow them to explore it further. Image analysis can be usefully employed at virtually all levels of morphology, from the macroscopic to the sub-microscopic domains. It remains relatively expensive in terms of contemporary biology budgets, but an investigator with a research grade microscope can enter the area for as little as about $22,000 to $27,000 today. There seems little doubt that some level of image analysis capacity should be available in most biology programs. Kaplow (1984) has recently reviewed image analysis options, and recognized some of the problems and options this technology offers. He, too, proposes a "small system" using a 16 bit microcomputer that could handle 512 X 512 pixel images digitized to at least 64 grey levels, a pair of dual-sided dual-density 5 1/4" floppy disks, a high-resolution RGB color monitor and a suitable dot matrix printer. It seems very probable that **these costs could be further eroded, and that various forms of manipulation of digitized images will pervade** biology at all levels.

LITERATURE CITED

Aherne WA, Dunill MS (1982) Morphometry. Edward Arnold, London.

Bacus JW (1974). Erythrocyte morphology and centrifuged "spinner" blood film preparations. J. Histochem. Cytochem 22:506.

Barrows GH, Sisken JE, Allegra JC, Grasch SD (1984). Measurement of fluorescence using digital integration of video images. J. Histochem. Cytochem. 32:741.

Bartels PH, Buch Roeder RA, Hillman DW, Jonas JA, Kessler D, Shoemaker RM, Shack RV, Tower D, Vukoratovitch D (1981). Ultrafast laser scanner microscope: Design and construction. Analyt. Quant. Cytol 3:55.

Boss JMN and Mays RGW (1976). Critical evaluation of histogram densitometry of Feulgen-stained normal human lymphocyte nuclei. J. Microscop. 108:277.

Boss JMN, Feneley RCL, Mays RGW (1976). Modes of chromatin packing in Feulgen-stained normal and malignant uroepithelial cells nuclei, and in lymphocytes, in relation to DNA content. J. Microscp. 108:291.

Cornelisse CJ, Ploem JS (1976). A new type of two-color fluorescence staining for cytology specimens. J. Histochem. Cytochem. 24:72-81.

Cowan WM, Wann DF (1973). A computer system for the measurement of cell and nuclear sizes. J. Microscop. 99:331.

Curtis SK, Cowden RR (1980). The effects of fixation and preparation on three quantitative fluorescent cytochemical procedures. Histochemistry 68:29.

Galbraith W. Marshall PN, Lee ES, Bacus JW (1979). Studies on Papanocolaou staining. I. Visible light spectra of stained cervical cells. Analyt. Quant. Cytol. 1:169.

Grinvald A, Anglister L, Freeman JA, Hilesheim R, Manker A (1984). Real-time imaging of naturally evoked electrical activity in intact frog brain. Nature (Lond.) 308:848.

Kaplow LS (1984). The Histochemist's phantom image analyser. Histochem. J. 16:617.

Leif RC, Easter HN, Warters RL, Thomas RL, Dunlap LA, Austen MF (1971). Centrifugal Cytology I. A quantitative technique for the preparation of glutaraldehyde-fixed cells for the light and scanning electron microscope. J. Histochem. Cytochem. 26:1018.

Lockart RZ, Pezzella KM, Kelley MM, Toy ST (1984). Features independent of stain intensity for evaluating Feulgen-stained cells. Analyt. Quant. Cytol. 6:105.

Marshall PN, Galbraith W (1984). Microspectrophotometric studies of Romanowsky stained blood cells. III. The action of methylene blue and Azure B. Stain Tech. 59:17.

Mathog D, Hochstrasser M, Greunbaum Y, Saumweber H, Sedat J (1984). Characteristic folding pattern of polytene chromosomes in Drosophila salivary gland nuclei. Nature (Lond.) 308:414.

Mayall BH, Mendelsohn ML (1970). Deoxynebonucleic acid cytophotometry CYDAC, the theory of scanning photometry and the magnitude of residual errors. J. Histochem. Cytochem. 18:383.

Mello MLS (1983). Cytochemical properties of euchromatin and heterochromatin. Histochem. J. 15:739.

Montgomery P.O'B, Bonner WA, Cook JE (1964). Flying and stepping spot television microscopy. J. Ray Mic Soc. 83:73.

Oud PS, Zahniser DJ, Garcia GL, vanBoekel MCG, Haag D,J, Hermkens HG, Pahlplatz MMM, Voojis PG, Herman CJ (1984). Pressure fixation method of transferring cells from polycarbonate filters to glass slides. Analyt. Quant. Cytol 6:71.

Piper J (1982). Interactive image enhancement and analysis of prometaphase chromosomes and their band patterns. Analyt. Quant. Cytol. 4:233.

Ruter A, Wittekind D, Gunzer U, Aus HM, Harms H (1982). Comparative colorimetry in cytophotometric measurements of azure B-eosin Y-stained and Gilmsa-stained blood all smears. Analyt. Quant. Cytol. 4:128.

Serra J, (1982). Image Analysis and Mathematical Morphology. Academic Press, New York.

Tanasugarn L, McNiel P, Reynalds GT, Taylor DG (1984). Microspectrofluorometry of digital image processing: measurement of cytoplasmic pH. J. Cell Biol. 98:717.

Tourassis VD, Dekker A, Preston K (1983). Relationship between cell size and weight of human liver; an automated morphometric study. Analyt. Quant. Cytol. 5:43.

Wied GL, Bartels PH, Dyten HI, Pishotta, Bibbo M (1982). Rapid high-resolution cytometry. Analyt. Quant. Cytol 4:257.

Weibel ER (1979). Stereological Methods. Vol 1. Practical Methods for Biological Morphometry. Academic Press, New York.

_____ (1980). Stereological Methods. Vol 2. Theoretical Foundations. Academic Press, New York.

Williams GZ (1964). Applications of the videcon and image converter in ultraviolet microscopy. J. Ray. Mic. Soc. 83:69

SCANNING OPTICAL MICROSCOPY

Tony Wilson

Department of Engineering Science,
University of Oxford,
Parks Road, Oxford, OX1 3PJ, United Kingdom.

The conventional microscope is an example of a parallel processing system, in which the whole area of the specimen is simultaneously imaged either onto a screen or directly onto the retina of the eye. While this can be quite adequate in many cases, the format of the image is not readily suited to subsequent electronic processing, nor can the optical system be easily adapted to take advantage of the various resolution enhancement schemes which we shall discuss later in this chapter.

A sequential imaging system provides a much more **versatile approach. It may be achieved by scanning a** diffraction limited spot of light relative to a specimen in a raster type scan. In this way the image is built up point by point, and may be displayed on a TV screen or stored in a computer for future processing. The first example of this kind of light microscope was reported in 1951 by Roberts and Young[1]. Their flying spot microscope, which was intended for biological studies, used a scanning spot of light from the face of a flying spot scanner tube. The light from the raster was transmitted through conventional microscope optics in reverse, producing a tiny spot of light which was scanned over the sample.

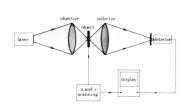

Figure 1.

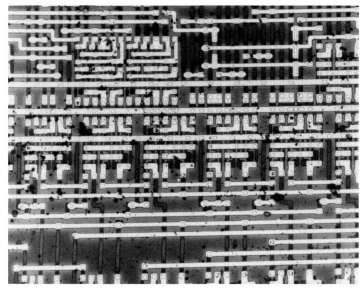

Figure 2.

Any radiation passing through the sample fell on a photodiode, where it was converted into an electrical signal. The signal was then appropriately amplified and used to modulate the intensity of a TV display scanned in synchronism. The later invention of the laser and its incorporation as the light source heralded the birth of the scanning optical microscope, Figure 1. There are essentially two different methods of scanning in these instruments. The alternatives are either to scan a focussed light beam across a stationary object, or to scan the object across a stationary spot. Mechanical scanning has the advantage that the optical system is very simple and that the lenses need only be corrected for on-axis aberrations. Figure 2 shows a high quality reflection image of an integrated circuit obtained with a mechanically scanned instrument. No contrast enhancement has been used on this image but the electrical form of the image allows electrical contrast enhancement to facilitate the observation of weak detail and precludes the need to stain biological specimens, thus eliminating the risk of killing or altering living cells in the staining process.

It has been shown[2,3] that there are two important optical configurations of scanning microscopes. The first, the Type 1 arrangement, has imaging properties identical to

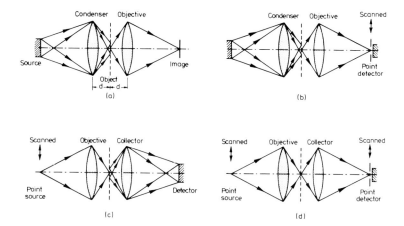

Figure 3.

the conventional microscope, while the other, known as the Type 2 or confocal, provides greatly improved imaging.

Figure 3(a) illustrates the optical system of a conventional microscope, in which the object is illuminated with a patch of light from an extended source through a condenser lens. The object is then imaged by the objective as shown, and the final image is viewed through an eyepiece. In this case the resolution is due primarily to the objective lens, while the aberrations of the condenser are unimportant. A scanning microscope using this arrangement could then be realised by scanning a point detector through the image plane so that it detects light from one small region of the object at a time, thus building up a picture of the object point by point, Figure 3(b). The arrangement of Figure 3(c), using a second point source and an incoherent detector has the same imaging properties. This is the arrangement of the Type 1 scanning microscope. The point source illuminates one very small region of the object, while the large area detector measures the power transmitted by the collector lens. The arrangement shown in Figure 3(d) is a combination of those in Figures 3(b) and 3(c). Here the point source illuminates one very small region of the object, and the point detector detects light only from the same area. An image is built up by scanning the source and detector in synchronism.

In this configuration we see that both lenses play equal parts in the imaging and are employed simultaneously to image the object, and hence the resolution is expected to be improved; and this prediction is bourne out by both calculation and experiment. This arrangement has been named a Type 2 or confocal scanning microscope. The term confocal is used to indicate that both lenses are focussed on the same point on the object. In practical arrangements, as we have already mentioned, it is often more convenient to scan the object rather than the source and detector together.

This resolution improvement may at first seem to contravene the basic limits of optical resolution. It may be explained, however, by a principle due to Lukosz$^{(4)}$ which states that resolution may be increased at expense of field of view. The field of view can be increased, however, by scanning. One way of taking advantage of Lukosz's principle is simply to place a very small aperture extremely close to the object. The resolution is now determined by the size of the hole rather than the radiation. This scheme has been successfully demonstrated at microwave frequencies$^{(5)}$, but there are such severe practical difficulties in locating a small enough aperture close enough to the object that the scheme has not been applied at optical frequencies.

This does not mean, however, that we cannot take advantage of the principle at optical frequencies. If, instead of using a physical aperture in the focal plane, we use the back projected image of a point detector in conjunction with the focussed point source, we have a confocal scanning microscope. This, then, is a practical arrangement which gives superior resolution at optical frequencies.

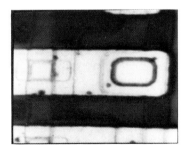

Figure 4(a)   Figure 4(b)

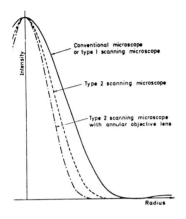 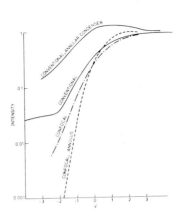

Figure 5   Figure 6

Figure 4(a), for example, shows a conventional image of a portion of a microcircuit, and Figure 4(b) shows the confocal image of the same area. The resolution improvement is apparent.

The confocal microscope behaves as a coherent optical system in which the image of a point object is given by the product of the image formed by the lens before the object (the objective) with that formed by the lens behind the object (the collector). This results in a sharpened image of a single point with extremely weak outer rings, which gives rise to images without artefacts. Figures 5 and 6 show theoretical images of a single point and straight edge object respectively. In both cases the confocal image is superior.

An inevitable consequence of the point detector in the confocal microscope is that the microscope images detail only from parts of the object around the focal plane. This depth discrimination property is not a feature of conventional microscopes. The effect may be explained physically by considering the light from outside the focal plane, Figure 7. The use of a central point detector results in a much weaker detected signal and so

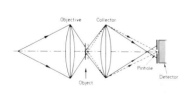

Figure 7

provides discrimination in the image against detail outside the focal plane.

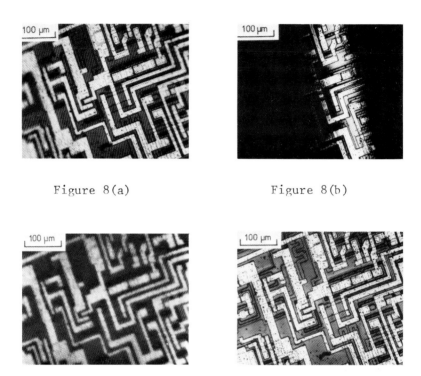

Figure 8(a)

Figure 8(b)

Figure 8(c)

Figure 8(d)

The effect has been illustrated by deliberately mounting a planar microcircuit in a scanning optical microscope with its normal at an angle to the optic axis. Figure 8(a) shows the Type 1 microscope image: only one portion, running diagonally, is in focus. Figure 8(b) is the corresponding confocal image: here the discrimination against detail outside the focal plane is apparent. The areas which were out of focus in Figure 8(a) have now been rejected. Furthermore the confocal image appears to be in focus throughout the visible band, demonstrating that the depth discrimination effect is dominant over the depth of focus. This is important, as it means that any detail that is imaged efficiently in a confocal microscope will be in focus.

It is now possible to take advantage of the depth discrimination property and combine it with axial scanning to produce a high resolution imaging instrument with vastly increased depth of field.

For example, if we had moved the specimen of Figure 8(a) and 8(b) axially, we would have obtained a corresponding pair of images, but with a different portion of the object brought into focus(6). Thus we can produce the image of a rough object such that all areas appear in focus by scanning the object in the axial direction with an amplitude sufficiently large for every part of its surface to pass through the focal plane.

Figures 8(c) and 8(d) show the effect of mounting the microcircuit specimen on a piezoelectric bimorph and scanning it axially through a distance of about 20 μm. It is clear that the conventional microscope gives the expected confused, blurred, out of focus image, whereas in the confocal mode only the in focus detail has been accepted by the microscope, thereby producing an image which appears in focus across the whole field of view, Figure 8(d). This should be compared particularly with the conventional image of Figure 8(a). The main merit of this approach is that by introducing axial scanning, we are able to produce 'projected' images of rough specimens with resolution comparable with that produced in a conventional microscope by a high numerical aperture lens.

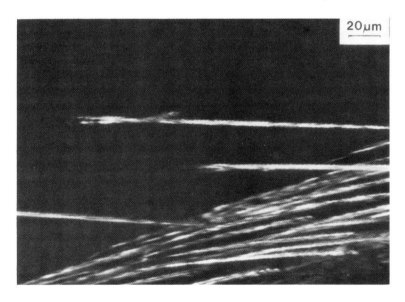

Figure 9

This technique is potentially extremely powerful, as it permits the depth of field in optical microscopy to be extended, in principle without limit, while still retaining high resolution, diffraction limited imaging. An experimental extension of more than two orders of magnitude has already been achieved [7]. An example of this technique at high resolution is shown in Figure 9 where the hairs on an ant's leg, with two hairs projecting to the left, have been imaged. The axial distance between the tips of these hairs is 30 μm, and Figure 9 is an extended focus image which show both excellently resolved along their full length, as well as much detail on the leg itself.

By using detectors with different sensitivity geometries, it is possible to produce images which depend on specific object properties. Dekkers and de Lang [8], for example, have proposed the use of a large area detector split into two halves to obtain a differential phase contrast image by displaying the difference signal from the two halves of the detector, Figure 10. Figures 11 and 12 show comparison images obtained by this method (11) and the conventional Nomarski technique (12), illustrating the superior resolution of the split detector mode.

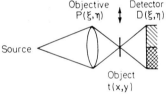

Figure 10

Figure 11

Figure 12

The use of scanning in microscopy allows the use of a wide range of imaging mechanisms which cannot be exploited in conventional microscopy. These modes generally rely on light input to produce some observed effect. An example is the optical beam induced contrast (OBIC) method for imaging electronic properties of semiconductor devices(10), in which the laser beam excites carriers to produce a current in an external circuit. Such an image shows up dislocations, grain boundaries, and other defects. Figure 13(a) is a reflected light image of a polycrystalline solar cell and Figure 13(b) is the corresponding OBIC image showing both surface and subsurface grain boundaries. Stacking faults in a silicon transistor are imaged by the OBIC method in Figure 14.

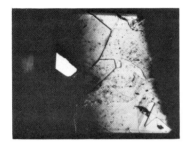
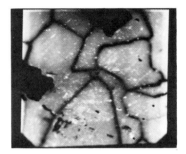

Figure 13(a)   Figure 13(b)

A popular method of obtaining similar images and information is to use the EBIC technique in the scanning electron microscope. The drawbacks with this method include specimen charging problems, which sometimes prevent an image being obtained, and specimen damage caused by the high energy

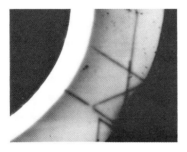
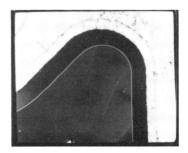

Figure 14   Figure 15

of the electron beam and contaminants within the vacuum column.
Figure 15 is a reflection scanning optical microscope image
of part of a silicon transistor which had previously been
examined in the scanning electron microscope. The damage to
the areas which had been scanned is clearly visible.

As a further example, Figure 16(a) shows an OBIC image
of a GaAs light emitting diode, while Figure 16(b) shows the
image of the light emitted by the diode itself[11]. The
emitted light image shows very clearly that this diode is not
uniformly efficient, but emits light only around the edge of
its active area. This is very useful information which
cannot easily be found from the OBIC image, and certainly
cannot be seen from a reflected light image.

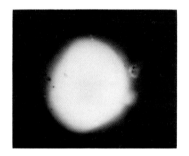
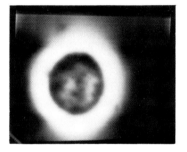

Figure 16(a)   Figure 16(b)

Finally the use of a laser allows the investigation of
the non-linear optical properties of an object. For example,
images have been formed from the second harmonic produced
within the object itself[12]. The quantity of harmonic
produced depends on the crystal structure and orientation of
the object. The harmonic microscope also exhibits a depth
discrimination property similar to that in the confocal
microscope. The detected signal is proportional to the
square of the input power, so that an appreciable amount of
harmonic is formed only in the region of focus. The
resolution in harmonic microscopy is increased[13] in relation to
that attainable with the primary frequency, as the
squaring sharpens the beam in the lateral direction. It is
important to note that because the harmonic radiation need
only be collected and not focussed, the smallest usable
wavelength is determined by the optics available for the

primary light. A wide range of other non-linear optical effects may also be used, including Raman scattering or two-photon fluorescence. In this way, information may be obtained about the energy levels and hence the chemical structure of the object.

REFERENCES.

1. J. Z. Young and F. Roberts, Nature, 167,231,(1951)
2. T. Wilson with C. J. R. Sheppard, Theory and Practice of Scanning Optical Microscopy, Academic Press, London, (1984)
3. C. J. R. Sheppard and T. Wilson. Optica Acta,25,315,(1978)
4. W. Lukosz, J. Opt. Soc. Am.,56,1563,(1966)
5. E. A. Ash and G. Nicholls, Nature, 237,510,(1972)
6. T. Wilson and D. K. Hamilton, T. Microsc,128,139,(1982)
7. C. J. R. Sheppard, D. K. Hamilton and I. J. Cox.
8. N. H. Dekkers and H. de Lang, Optik,41,452,(1974)
9. D. K. Hamilton and C. J. R. Sheppard, J. Micrsoc. (1984)
10. T. Wilson, W. Osicki, J. N. Gannaway and G. R. Booker, J. Mater. Sci.,14,961,(1979)
11. T. Wilson, J. N. Gannaway and P. Johnson, J. Micrcosc., 118,309,(1980)
12. J. N. Gannaway and T. Wilson, Proc. Roy. Microsc. Soc., 14,170, (1979)
13. T. Wilson and C. J. R. Sheppard, Optica Acta.,26,761,(1979)

A NEW INSTRUMENT COMBINING SIMULTANEOUS LIGHT AND SCANNING ELECTRON MICROSCOPY

Cornelia H. Wouters, Johan S. Ploem

Department of Histochemistry and Cytochemistry,
University of Leiden, Wassenaarseweg 72,
2333 AL Leiden, The Netherlands

INTRODUCTION

When individual cells are studied with a scanning electron microscope in the secondary emissive mode, the image produced will have a relatively great depth of focus and a high resolution, allowing the study of fine surface features. However, this secondary emissive mode reveals no information concerning the internal cell structures (e.g. nuclear chromatin patterns and cytoplasmic granules). To understand the structure and function of cells (especially in a mixed population), the correlation between morphological, physical, and cytochemical properties must be investigated. Wetzel et al. (1973) pointed out that the identification of individual cells with a scanning electron microscope alone is sometimes impossible.

With scanning electron microscopy (SEM) a narrow beam of electrons is produced in an electron gun at one end of the vacuum column and is focussed to a small spot on the surface of a specimen placed at the far end of the column. A probe (10 nm or less) is moved over the specimen in a regular pattern. The incident electrons (primary electrons) fall onto the specimen with a high velocity, impinge with the atoms of the specimen and give rise to several types of radiation. The most commonly employed mode of a scanning electron microscope uses the secondary electrons, which are produced by inelastic collisions between the primary electrons and the electrons in the target area of the specimen.

Light microscopy (LM) on the other hand gives information concerning the internal cell structure and the presence of special macromolecules after appropriate cytochemical staining. For this reason several approaches for correlating LM and SEM have been proposed in the last years.

McDonald et al. (1967) were among the first to use this correlation on microtome sections. SEM examination was performed on stained paraffin sections, which were coated with a platinum-palladium alloy. Afterwards the preparations (made on cover-glasses) were mounted on glass slides and studied with the light microscope. A similar type of investigation was performed on isolated cells (McDonald and Hayes, 1969). Like many other methods where SEM and LM are combined, this implied transfering a specimen from a light microscope to a scanning electron microscope (Geissinger et al., 1973) or the reverse (McDonald and Hayes, 1969).

When the light microscope and scanning electron microscope used are different instruments an accurate relocation of the same cells is necessary. Relocation has been accomplished in many ways. A frequently used method consists of accurate re-positioning of the specimen based on identical coordinates of the two microscope stages. For example, Geissinger (1974) constructed a module for a SEM stage which could carry microscope slides of standard dimensions. Besides Geissinger, there were other groups who also used coordinates for relocation (McDonald and Hayes, 1969). Another relocation method consists of markers on the supporting glass. Several kinds of markers have been employed, varying from a gold shadow produced by gold coating of a grid, which was placed on a cover-glass and removed afterwards (Michaelis et al.,1971), or India ink (Wetzel et al.,1973) to etching (Bahr et al., 1976) or printing a pattern on the supporting glass (Thornthwaite et al., 1976). Lytton et al. (1978) used the bleaching effect of the electron beam of the scanning electron microscope on a Millipore filter. Sometimes photographs were used as maps (Becker et al., 1981).

Shelton and Orenstein (1975) marked the supporting Millipore filter. An extensive survey on this subject is given by Wouters et al. (in press).

A disadvantage of all these methods is that they are rather time-consuming, with limited possibilities for alternating the light microscope and the scanning electron microscope for re-investigations. When interesting cells had been individually marked (van Ewijk and Mulder, 1976), only a limited number of cells could be investigated with LM and SEM for reasons of efficiency. Another disadvantage consists of the removal of the cover-glass, that is in many cases used for LM examination. Cells can be lost or damaged by the removal of the cover-glass.

To circumvent all these problems caused by using two separate microscopes, an instrument has been developed for the combined use of SEM and LM (Hartmann et al., 1978; Ploem, 1981; Ploem and Thaer, 1981). The attachment of a microfluorometer unit (MPV-II, Leitz, W Germany) enables measurement of fluorescence signals after an appropriate staining of cells in this combined LM/SEM microscope (Wouters and Koerten, 1982).

DESCRIPTION OF THE COMBINED INSTRUMENT

The instrument (LM/SEM) which combines SEM with LM, consists of a Leitz 1200 B scanning electron microscope with special adaptations, connected to a slightly changed Leitz light microscope. In Fig.1 this instrument is shown.

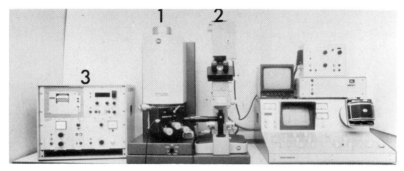

Fig.1. The instrument (LM/SEM) combines a scanning electron microscope (1) and a light microscope (2) by an intermediate optical system. Connected to this LM/SEM is a photometer unit (3).

The adaptation which makes the SEM part of this instrument different from the other scanning electron microscopes is the special stage in the vacuum chamber.

This stage, visible in Fig.2, incorporates LM optics and is connected via an optical bridge to the light microscope.

Fig.2. The special specimen stage, showing the LM optics (objective (1), condenser (2)) and the specimen in its holder (3).

The objective and condenser are built onto the stage inside the vacuum chamber. A photometer unit (Leitz MPV-II, W Germany) is attached to this light microscope. Also connected to the LM part of the LM/SEM is a camera, which makes it possible to take micrographs of the specimens. Fig.3 schematically illustrates the basic arrangement around the specimen to be investigated.

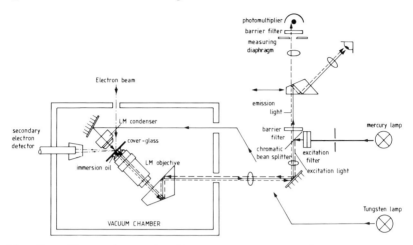

Fig.3. The schematic LM arrangement in the vacuum chamber. The path of the electron beam is visible (primary and secondary electrons) and the two light paths for transmitted illumination (absorption) and epi-illumination (fluorescence) are shown.

An exchangeable oil-immersion objective and a specially modified medium power objective with a long free working distance (serving as a condenser) are mounted inside the vacuum chamber attached to the specimen stage. Specimens prepared on cover-glasses (described below) are placed above the oil-immersion objective (N.A. 1.30), which is orientated with its optical axis at an angle of 135 degrees to the electron beam and 90 degrees to the cover-glass. A vacuum-resistant fluorescent free immersion oil is applied. Above the cover-glass the condenser is positioned in the same optical axis as that of the first objective. The condenser can be swung out of the light path of the optical axis to permit completely undisturbed electron-beam passage.

The specimen is inclined 45 degrees with respect to the electron beam and can be moved in the X and Y directions (up to 6 mm). The objective (under the specimen) can also be moved, first of all in the Z direction (for focusing) and secondly in the X and

Y directions (since the LM image has to be obtained from the place at which the electron beam hits the specimen). These movements are controllable from outside the vacuum chamber. Fig.4 shows the front panel of this vacuum chamber with the centering and focusing devices for the objective and the handle for swinging the condenser in and out the light path.

Fig. 4. The front panel of the vacuum chamber. Visible are the focusing device (1) and the centering device (2) for the objective, the devices for X and Y movement of the specimen table (3) and the condenser handle.

In Fig.3 the path of the primary and secondary electrons is visible. Furthermore two light paths are illustrated. The first light path is for transmitted illumination. The external light is produced by a tungsten lamp (attached to the LM part of the LM/SEM) and is guided by a glassfibre cable to the vacuum chamber. Here the light is led by mirrors and a prism inside the chamber to the condenser. After traversing the condenser, the light illuminates the specimen and is collected by the oil-immersion objective. From here the light leaves the vacuum chamber through the optical bridge and enters the LM part of the LM/SEM. It can now either be studied through the eyepieces or micrographs can be taken. Secondly, for fluorescence microscopy light is obtained from a mercury high-pressure lamp (again attached to the LM part of the LM/SEM). The light passes through the optical bridge into the vacuum chamber. Here the exciting light is concentrated by the oil-immersion objective onto the specimen (epi-illumination). The emitted fluorescence light returns to the LM part of the LM/SEM by the same path. It can again either be studied through the eyepieces or photographs can be taken. The attachment of a microfluorometer unit to the LM part of the instrument provided with a fluorescence illuminator (Ploem, 1969) also allows measurement of the intensity of the emitted light from a certain object (e.g. the quantitatively stained nucleus of a cell).

SPECIMEN PREPARATION FOR LM/SEM EXAMINATION

Since two microscopical techniques are combined in one instrument, the specimens which are examined in this instrument have to fulfill the general combined requirements for SEM and LM.

For SEM biological objects (e.g. cells or tissue) which are going to be investigated have to be fixed and the water in the object has to be replaced by a polar organic solvent. Subsequently the objects have to be dried, as they must be placed inside vacuum chamber. Furthermore, to avoid irregular charging of the specimen surface which can seriously and unpredictably affect the secondary electrons, the whole specimen has to be made conductive. To accomplish this, specimens are usually coated with gold or other conductive materials.

The most important requirement for the preparations made for the combined instrument is that they have to be made on cover-glasses instead of on stubs, mostly used to mount specimens for SEM examination. Transparency is essential for the LM part.

In addition the preparation has to be conductive. To achieve this, the cover-glass is glued with its corners to an aluminum holder with a hole in the middle (diameter 20 mm) so that the objective can focus later on the object. Conductive carbon cement is used as glue. The specimen is coated with gold by means of a sputter coater.

If cells from a suspension are to be examined with the LM/SEM, they are brought onto a cover-glass. This can be achieved by one of several methods. For instance, cells which have the capacity of active attachment (e.g. macrophages) will attach to glass before fixation is performed (Linthicum et al., 1974; Sanders et al., 1975). This has been described as the sedimentation method. However, fine details of the cell surface might be influenced by large variations in cell size due to the flattening of certain cell types brought on glass in an unfixed state. For this reason it is also possible that cells fixed prior to attachment are examined. But then their attachment will be less tight. The same is the case for cells that do not or hardly actively attach to glass, even in an unfixed state. Coating of the cover-glass with e.g. poly-L-lysine will promote cell adhesion, but many cells will still be lost (low recovery) under conditions of unit gravity sedimentation. Moreover such methods are very time-consuming (Sanders et al., 1975).

Wouters et al. (1974) described a preparation method in which fixed cells (or non-adherent cells) can be brought onto cover-glasses for LM/SEM examination. This method takes just a few minutes, and even after cytochemical staining the recovery is high. Prefixed cells are centrifuged on poly-L-lysine coated cover-glasses using the Leiden centrifugation bucket (van Driel-Kulker et al., 1980). This bucket (Fig.5) consists of two parts: a steel holder that fits into the centrifuge and a plastic inner part containing two cylindrical holes with a diameter of 1.5 cm each.

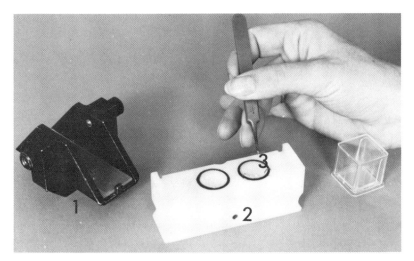

Fig.5. The centrifugation bucket used for the preparation of fixed cells for the combined LM/SEM microscope. The bucket consists of two parts: a steel holder (1) and a polyethylene inner part (2) containing two cylindrical holes (1.5 cm diameter). Poly-L-lysine coated cover-glasses (3) are placed onto the wet rubber rings.

Before assembling, two cover-glasses (1.8 x 1.8 cm) are placed on the lower side of the holes onto the wetted rubber rings (preventing leakage). Then, the inner part of the bucket with the cover-glasses is placed on a slide inside the steel holder. A suspension of fixed cells is placed in each hole of the bucket. Centrifugation is then performed. The centrifugation time and force depends on the cell type.

For peripheral blood mononuclear cells a centrifugal force of 70 g over a period of 10 min is sufficient for a high recovery. For fixed cervical epithelial cells 1200 g over a period of 15 min is necessary. Disassembling is done under liquid to prevent premature drying.

The fixation (either before or after bringing the cells onto glass) also depends on the cell type. For peripheral blood mononuclear cells 0.1% glutaraldehyde and 1% paraformaldehyde in 0.1 M cacodylate buffer is used for 30 min. The cervical epithelial cells are fixed in suspension with carbowax (2% (g/w) polyethylene glycol, MW 1.500, in 50% ethanol). Then the cells are placed on cover-glass by using the centrifugation bucket and post-fixed with 100% ethanol for 30 min. For a more detailed description of the preparation of cervical cells, the reader is referred to the paper by van Driel et al. (1980).

The cells are quantitatively stained for DNA with a Feulgen acriflavine-SO2 staining (Tanke and van Ingen, 1980) with the following modification. The whole staining procedure is performed at room-temperature in a small petridish. The Feulgen hydrolysis is performed 15 min or 30 min (respectively for peripheral blood mononuclear cells or cervical epithelial cells) in 5.0 N hydrochloric acid. The hydrolysis is stopped by rinsing 3 min in demineralized water. Subsequently the preparations are stained in a solution of 0.01% (w/v) acriflavine-SO2 (Chroma, W Germany) in 14 mM hydrochloric acid and 6 mM potassium metabisulphite for 15 min, and rinsed 6 min in demineralized water. In order to remove non-covalently bound dye, the petridishes are filled with acid ethanol (1% hydrochloric acid w/v in 70% ethanol) for 10 min. Thereafter the preparations are washed first with demineralized water, then with phosphate buffer, pH 6.8 and finally again with demineralized water (5 min each). This is followed by a dehydration in ethanol 70%, ethanol 80%, ethanol 90% during 5 min each, ethanol 100% (twice) 15 min each and ethanol 100% 30 min.

The preparations are critical point dried with CO2, glued with conductive carbon cement (Neubauer, W Germany) to special aluminum holders with a round opening in the middle (diameter 2 cm) and coated with gold. These preparations are investigated in the combined microscope.

APPLICATIONS

Several applications are possible with the combined LM/SEM instrument. First of all the general problem of specimen location is easily solved. Take for instance a tissue section examined in the combined microscope. Cytochemical staining will reveal the position of nuclei or other cell organelles, which makes the location of these cells in the tissue (after LM examination) in the scanning electron microscope simple. Also when a cell suspension is examined with the LM/SEM instrument, cells can be examined without losing time for relocation or moving the specimen from one microscope to the other.

Secondly recognition of several cell types is simplified. After an appropriate cytochemical staining the shape of the nucleus can help in identifying the cell type.

A third application is the possibility of combining the cell surface morphology with cytochemical markers of cancer cells. The cancer cells (e.g. from the cervix) are identified on the basis of their LM morphology and DNA content (measured with the microfluorometer) after a quantitative cytochemical staining. With the SEM part of the combined instrument their surface morphology can then be studied.

Several examples of the applications are given below. The preparation method including the DNA staining and the critical point drying has been tested on human peripheral blood mononuclear cells (PBMC). These cells were attached and positioned on a cover-glass and prepared as described in the specimen preparation part, the fixed cells were centrifugated onto glass, stained with acriflavine-Feulgen, dehydrated and critical point dried. The results of the examination in the combined instrument are shown in Fig.6. Fig.6a shows the absorption LM image and the fluorescence LM image of a group of mononuclear blood cells (lymphocytes and monocytes). The SEM image is shown in Fig.6b. With the absorption image (Fig.6a) a lymphocyte is located, focused upon and the measuring diaphragm is placed around the nucleus. A

DNA measurement can then be performed using the fluorescence signal of the nucleus (Fig.6a). Under completely dry conditions and covered with gold the DNA content of fifty lymphocytes was measured and the SEM image was studied. Measurements show a diploid peak with a coefficient of variation of 6.1%, which is acceptable for this kind of application. The SEM images of these cells reveal the normal surface morphology of lymphocytes investigated with a scanning electron microscope. So, although the cells had been quantitatively stained, the SEM image remains comparable with cells which had not been cytochemically stained for LM but only processed for SEM examination. This experiment indicates that it is possible to study with the LM/SEM the correlation between cell cycle (DNA measurements) and cell surface morphology (SEM).

To investigate the possibilities of the LM/SEM for the analysis of epithelial cancer cells, a pilot study has been performed on normal cervical epithelial cells. These cells were centrifuged onto a cover-glass and prepared as described above. In Fig.7a the absorption LM image and the fluorescence LM image is shown. The SEM image of the same cells is visible in Fig.7b. These cells can also be measured. In this case a normal mature intermediate cell is shown with a 2N DNA content and a normal surface morphology (a regular pattern of microridges). In spite of the quantitative DNA staining, the SEM morphology remains the same as described in the literature for normal SEM preparations of these cells (Lanes et al.,1973; Williams et al.,1973). The nucleus is slightly visible because it protrudes through the cytoplasmic surface. Lines on the cell (called crests or bars) are visible indicating that the luminal side is shown (Jordan and Williams, 1971). Shallow, branching and interlacing convolutions are present, forming a system of microridges which cover the entire cell surface. As a rule no particular orientation of the microridges is observable in the central part of the cell. At the periphery some orientation parallel to the cell boundary is occasionally seen. Research with the LM/SEM is in progress to investigate the relationship between the DNA content of malignant cervical cells and their surface morphology.

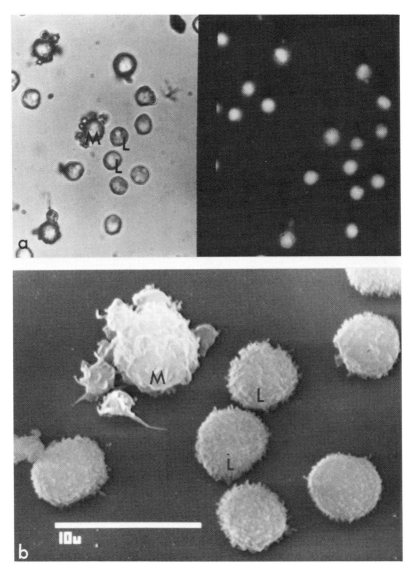

Fig.6. Combined LM/SEM micrographs of human lymphocytes (L) and monocytes (M) attached to glass by centrifugation after fixation and quantitatively stained for DNA with acriflavine-Feulgen and prepared for SEM. a. LM absorption image and LM fluorescence image. b. SEM image of the same group of cells.

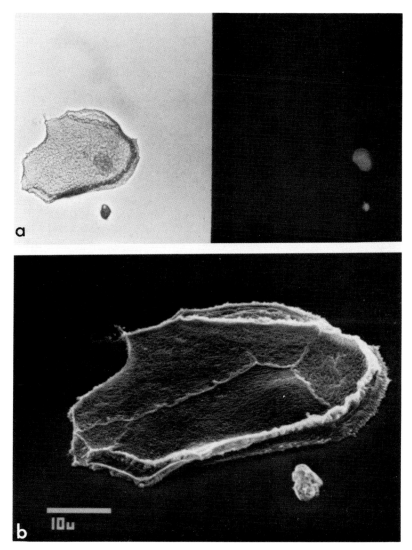

Fig.7. Combined LM/SEM micrographs of a normal human cervical epithelial cell attached to glass by centrifugation after a weak fixation, stained for DNA with acriflavine-Feulgen and prepared for SEM. a. LM absorption image and LM fluorescence image. b. SEM image of the same cell showing a regular pattern of microridges.

Also in progress is the investigation of a fourth application, the use of immunocytochemistry with this microscope. Special attention will be given to the detection of small gold markers with the reflection contrast mode of the LM/SEM. Since these gold markers are below the resolving power of the scanning electron microscope they will not disturb the secondary electron image. On the other hand, they will reflect light and because of this reflection they will be visible with the LM part of this instrument. Until now immunocytochemistry in a scanning electron microscope has only been used with different surface labels. For example, the antibody has been linked to latex spheres (Lobuglio et al., 1972; Molday et al., 1974) or silica spheres (Peters et al., 1976). Other markers like hemocyanin, TMV, ferritin, F2 and T2 bacteriophages and larger gold particles have also been used. All of them were visible in the SEM image. These markers will, although they provide a way to detect the positive cells in the scanning electron microscope, disturb the image. A virus as a marker may be confused with cell surface projections like microvilli. Due to the high sensitivity of the reflection contrast part of the LM/SEM the above mentioned method with the small gold markers might prove to be a good alternative.

The applications above all consist of studies on human or animal cells, but the LM/SEM can also be used in botany or geology research, for example in the palyonological age determinations. In general it is assumed that the vegetation of a sedimentation area is reflected in the pollen content of a deposit, provided the circumstances were favourable for their fossilization. The vegetational changes will then be reflected in the changing pollen composition by which zones in a pollen diagram can be distinguished. The critical features of pollen and spores are that they possess exceptionally tough organic outer cell walls (exines), which are highly resitent to temperature, pressure, and non-oxidative chemical treatments (Sengupta and Muir, 1977). The pollen can be recognized in the scanning electron microscope according to their cell walls (Peat, 1981). However, sometimes the autochthonous

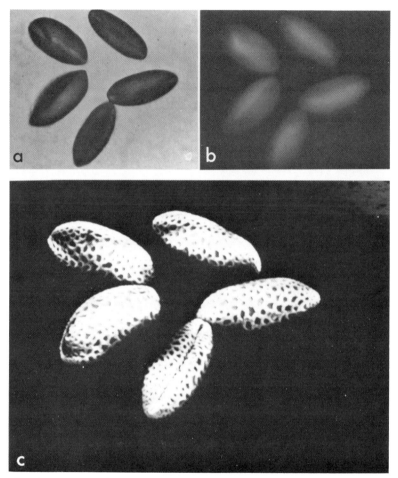

Fig.8. Combined LM/SEM micrographs of fresh pollen from Frittilaria meleagris glued onto a cover-glass with fluoromount. a. LM absorption image. b. LM fluorescence image showing the autofluorescence. c. SEM image of the same pollen showing the cell wall structure.

vegetation can be contaminated with re-worked pollen grains (secondary). An extra quality of these cell walls, their autofluorescence, will then help the determination of pollen (van Gijzel, 1967). Living pollen contain protoplasm (intine), the bright fluorescence of which is dominant over that of the exine. But during fossilization the intine disappears and only the exine is fluorescent. So each type of secondary pollen will show a larger variation in fluorescence colour than the autochthonous material. With increasing geological age there is a change in the fluorescence colour from blue or green to red or brown, due to the increase of pressure and/or temperature with increasing depth of burial. So besides the morphology of the cell wall as can be seen with the scanning electron microscope, the fluorescence of the cell wall helps to determine the fossil palynomorphs. An example of fresh pollen is given in Fig.8.

REFERENCES

Bahr GF, Bibbo M, Mikel U, Engler W, Rao C, Wied GL (1976). Correlation of light and scanning electron microscopy, a new method for exfoliative cytology. Acta Cytol 20:239.

Becker SN, Wong JY, Marchiondo AA, Davis CP (1981). Scanning electron microscopy of alcohol-fixed cytopathology specimens. Acta Cytol 25:578.

Driel-Kulker AMJ van, Ploem-Zaaijer JJ, Zwan-van der Zwan M van der, Tanke HJ (1980). A preparative technique for exfoliated and aspirated cells showing different staining procedures. Anal Quant Cytol 2:243.

Ewijk W van, Mulder MP (1976). A new preparation method for scanning electron microscopic studies of single selected cells cultured on a plastic film. IITRI/SEM 2:131.

Geissinger HD (1974). A precise stage arrangement for correlative microscopy for specimens mounted on glass slides, stubs or EM grids. J Microsc 100:113.

Gijzel P van (1967). Autofluorescence of fossil pollen and spores with special references to age determination and coalification. Leidse Geol Med 40:263.

Hartmann H, Hund A, Moll SH, Thaer A (1978). Attachment for combined scanning electron and light microscopical examinations. Beitr elektronenmikroskop Direktabb Oberfl 11:381.

Jordan JA, Williams AE (1971). Scanning electron microscopy in the study of cervical neoplasia. J Obst Gyn Brit Comm 78:940.

Lanes AT, Farre CB, Ferenczy A, Richart RM (1973). Scanning electron microscopy of normal exfoliated squamous cervical cells. Acta Cytol 17:507.

Linthicum DS, Sell S, Wagner RM, Trefts P (1974). Scanning immunoelectron microscopy of mouse B and T lymphocytes. Nature 252:173.

Lobuglio AF, Rinehart JJ, Balcerzak SP (1972). A new immunologic marker for scanning electron microscopy. IITRI/SEM:313.

Lytton DG, Yuen E, Rickard KA (1978). Scanning electron and light microscope correlation of individual human bone marrow cells before and after culture in nutrient agar. J Microsc 115:35.

McDonald LW, Hayes TL (1969). Correlation of scanning electron microscope and light microscope images of individual cells in human blood and blood clots. Exp and Mol Path 10:186.

McDonald LW, Pease RFW, Hayes TL (1967). Scanning electron microscopy of sectioned tissue. Lab Invest 16:532.

Michaelis TW, Larrimer NR, Metz EN, Balcerzak, SP (1971). Surface morphology of human leukocytes. Blood 37:23.

Molday RS, Dreyer WJ, Rembaum A, Yen SPS (1974). Latex spheres as markers for studies of cell surface receptors by scanning electron microscopy. Nature 249:81.

Peat CJ (1981). Comparative light microscopy, scanning electron microscopy and transmission electron microscopy of selected organic walled microfossils. J Microsc 122:287.

Peters KR, Gschwender HH, Haller W, Rutter G (1976). Utilization of high resolution spherical marker for labeling of virus antigens at the cell membrane in conventional scanning electron microscopy. IITRI/SEM:75.

Ploem JS (1969). The use of a cervical illuminator with interchangeable dichroic mirrors for fluorescence microscopy with incident light. Z wiss Mikrosk u mikrosk Tech 68:129.

Ploem JS (1981). A new instrument permitting simultaneous scanning electron microscopy and fluorescence microscopy of the same specimen by integrating a LM in a SEM vacuum chamber. Cytometry 2:121.

Ploem JS, Thaer A (1981). A light microscope integrated in the vacuum chamber of a SEM. Proc RMS 16:253.

Sanders SK, Alexander EL, Braylan RC (1975). A high-yield technique for preparing cells fixed in suspension for scanning electron microscopy. J Cell Biol 67:476.

Sengupta S, Muir MD (1977). The use of light and electron microscopy in the study of experimentally altered spores and pollen grains. J Microsc 109:153.

Shelton A, Orenstein JM (1975). A plating method for preparation of cells for culture and for observation by light or electron microscopy. Exp Mol Pathol 23:220.

Tanke HJ, van Ingen EM (1980). A reliable Feulgen acriflavine-SO2 staining procedure for quantitative DNA measurements. J Histochem Cytochem 28:1007.

Thornthwaite JT, Cayer ML, Cameron BF, Leif SB, Leif RC (1976). A technique for combined light and scanning electron microscopy of cells. IITRI/SEM 5:127.

Wetzel B, Erickson BW,Jr, Levis WR (1973). The need for positive identification of leukocytes examined by SEM. IITRI/SEM 3:535.

Wilding LP, Geissinger HD (1973). Correlative light optical and scanning electron microscopy of minerals: a methodology study. J Sedim Petrol 43:280.

Williams AE, Jordan JA, Murphy JF, Allen JM (1973). The surface ultrastructure of normal and abnormal cervical epithelia. IITRI/SEM:597.

Wouters CH, Hesseling S, Daems WTh, Ploem JS (1984). A centrifugation method for standardized sedimentation of mononuclear human blood cells on glass for scanning electron microscopy. J Microsc 136:315.

Wouters CH, Koerten HK (1982). Combined light microscope and scanning electron microscope, a new instrument for cell biology. Cell Biol Int Rep 6:955.

LIGHT MICROSCOPIC AND ULTRASTRUCTURAL APPROACHES TO THE INVESTIGATION OF CELL NUCLEI AND CHROMATIN

DNA "STANDARDS" AND THE RANGE OF ACCURATE DNA ESTIMATES BY FEULGEN ABSORPTION MICROSPECTROPHOTOMETRY

Ellen M. Rasch, Ph.D.

Department of Biophysics
Quillen-Dishner College of Medicine
East Tennessee State University
Johnson City, TN 37614

Many methods are available for the microscopic analysis of deoxyribonucleic acid (DNA) in tissue preparations. In particular, absorption cytophotometry of the Feulgen-DNA dye complex has been used in laboratories around the world for more than 30 years to determine relative amounts of DNA per nucleus and to monitor the quantitative behavior of DNA during normal and neoplastic cell growth and differentiation (Ris, Mirsky 1949; Swift 1950a; Swift 1950b; Pollister 1952; Swift 1953; Atkin, Richards 1956; Rasch et al 1959; Rasch, Woodard 1959; Atkin 1969; Atkin 1976; Cavanagh 1979; Sandritter 1979; Clark, Kasten 1983; Chieco, Boor 1984). More recently, DNA-Feulgen cytophotometry has been used to estimate genome size in a number of organisms of special interest to molecular biologists, developmental geneticists and evolutionary biologists who need a reliable estimate of the amount of DNA in the haploid genome of cell systems that are not amenable to direct biochemical analysis (Rasch et al 1959; McLeish, Sunderland 1961; Rasch et al 1965; Miksche 1968; Rasch, Barr 1969; Rasch et al 1970; Rasch et al 1971; Rees, Jones 1972; Bennett 1972; Cullis 1973; Price et al 1973; Thomson et al 1973; White 1973; Rasch 1974; Rasch 1976; Bennett, Smith 1976; Rasch et al 1977; Mulligan, Rasch 1980; Rasch 1980; Rasch, Rasch 1981; Rasch et al 1982).

The chemical specificity and quantitative validity of the Feulgen reaction have been empirically documented over the years by a number of investigators (Pollister, Swift 1950; Swift 1950; Swift 1953; Swift, Rasch 1954; Vendrely, Vendrely 1956; James 1965; Atkin et al 1965; Vaughn, Locy 1969; Bachmann 1972; Rasch 1974; Duijndam, van Duijn 1975;

Bedi, Goldstein 1976; Dhillon et al 1977; Kjellstrand 1980; Allison et al 1981; Pogany et al 1981; Clark, Kasten 1983). In practice, individual nuclei in preparations stained by the Feulgen reaction are measured with any one of a number of different kinds of cytophotometers (Altman 1975) and the relative amount of DNA per cell is reported in arbitrary Feulgen absorbance units, or so-called "C" values, which are DNA amounts associated with the haploid, diploid and/or polyploid chromosome complements of various classes of interphase nuclei from germ line or somatic cell populations (Swift 1950a; Swift, Kleinfeld 1953; Rasch et al 1959; Rasch 1970; Rasch 1974; Rasch et al 1977; Mulligan, Rasch 1980).

It is now quite common to express cytophotometric data based on amounts of Feulgen staining for an "unknown" type of cell in terms of an estimate of the absolute DNA content in picograms (pg) per nucleus or per haploid (1C) genome. Such estimates require the use of reference standards of known or presumed DNA content so that a series of relative values obtained in arbitrary photometric units of total absorbance of the Feulgen-DNA dye complex at a given wavelength can be translated into pg or other units, such as daltons or number of nucleotide pairs or kilobases (kb), that are more meaningful to biochemists and molecular biologists (Rasch, Woodard 1959; Rasch et al 1971; Rasch 1974; Dhillon et al 1977; Rasch et al 1977; Mulligan, Rasch 1980).

As pointed out by workers who have cautioned against the transformation of Feulgen-DNA absorbance values into actual DNA amounts (Swift 1950; Atkin et al 1965) there are inherent risks, as well as a residual uncertainty of some 5-10%, when photometric data are used to estimate absolute amounts of DNA (Bachmann 1972; Rasch 1974; Bennett, Smith 1976). However, as is the case with any biochemical assay for determining protein or nucleic acid levels by the use of a calibration curve of varying concentrations of a "known" reference standard, it is important to remember that cytophotometric estimates of absolute DNA amounts actually refer to units of chicken DNA, or _Xenopus_ DNA or to the particular type of DNA from whatever reference cells were used in establishing the calibration curve (Vaughn and Locy 1969; Rasch 1974). Such estimates also presume an evaluation of variables in tissue fixation, processing, preparation of the Schiff reagent and staining protocols, as well as careful attention to the many potential sources of systematic errors in measuring procedures (Swift 1950; Swift et al 1969;

Mayall and Mendelsohn 1970; Bachmann 1970; Duijndam, van Duijn 1975; Goldstein 1970; Goldstein 1971; Bedi and Goldstein 1976; Piller 1977; Miksche et al 1979; Duijndam et al 1980; Allison et al 1981; Sklarew 1983).

Although caution is advisable when making estimates of absolute amounts of DNA, there should be no ambiguity in interpreting cytophotometric data expressed as pg of DNA, so long as the relationships in question are explicitly stated in the form of photometric values expressed in relative, arbitrary units and their estimated equivalencies in terms of absolute amounts of DNA (Rasch et al 1970; Rasch et al 1971; Bennett 1972; Bachmann, Rheinsmith 1973; Rasch 1974; Bennett, Smith 1976; Rasch et al 1977).

Some workers prefer to express cytophotometric data for an unknown cell type as a percentage of the genome size of a given reference standard such as the diploid human genome, which generally is assumed to be 7 pg DNA (Atkin et al 1965; Ohno, Atkin 1966; James 1972; Bachmann 1972). Often the blood cell nuclei of chicken (2.5 pg DNA), trout (5.2 pg DNA) or Xenopus laevis (6.3 pg DNA) are used as cytophotometric reference standards (Rasch et al 1971; Rasch 1974; Thiebaud, Fischberg 1977; Coulson et al 1977; Coulson, Tyndall 1978; Vindelov et al 1980; Lee et al 1984). Bachmann (1972) has also used toad (Bufo) and mouse nuclei as reference standards of known DNA content. As shown by Rasch (1974), there is a reasonably direct correspondence between cytophotometric estimates of absolute DNA levels in a wide variety of cell types and the DNA content determined for these same kinds of cells by direct biochemical analyses. Thus, for ranges of DNA content from roughly 0.05 pg to 15 pg the stoichiometry of the reaction and the linearity in response of cytophotometers seem fairly well established.

There are several types of cell systems, such as the recently discovered polytene chromosomes in the pseudonurse cells of certain female-sterile, ovarian tumor mutants of Drosophila melanogaster (King et al 1981) that offer an interesting challenge for analysis by Feulgen cytophotometry because the DNA levels of these cells far exceed the range usually measured. Furthermore, the number of egg chambers at different stages of development within a given ovariole and the heterogeneity of follicle cell and pseudonurse cell nuclei within a single egg chamber preclude analysis of this system by conventional biochemical procedures. Quantitative cytochemical study of these nuclear populations uniquely requires the use of static cytophotometry so that individual types of cells can be identified unequivocally and measured.

The methodological studies to be reported here were prompted by our need to analyze the DNA levels associated with characteristic patterns of change in the morphology of the polytene chromosomes that accompany ovarian development in these unusual mutants of Drosophila (Rasch et al 1984). Simply put, to verify differences in DNA levels of chromatin aggregates that contain upwards of 200 to 800 pg DNA, it was necessary to determine the upper ranges over which DNA-Feulgen absorption microspectrophotometry can provide valid estimates of DNA content.

By the use of a variety of cell sources ranging from sperm of Drosophila melanogaster (0.18 pg DNA; Rasch et al 1971) to blood cell nuclei of the Congo eel; Amphiuma means, (almost 200 pg DNA, Conger, Clinton 1973)) and by varying the number of such objects within a given field scanned by an integrating microdensitometer, we found that estimates of DNA content can be obtained reliably over five orders of magnitude--from less than 0.2 pg to more than 2 ng of DNA.

Materials and Methods

We have used a variety of amphibian blood cell nuclei in addition to chicken, molly fish, carp and rainbow trout erythrocytes smeared in thin films that were treated in an identical manner with regard to each step in fixation, acid hydrolysis, staining and preparation for cytophotometry.

DNA reference controls were chosen from animals having erythrocyte nuclei with 1.6 pg to 193 pg DNA as determined from the literature (Mirsky, Ris 1951; Dawid 1965; Rasch et al 1970; Bachmann 1970; Shapiro 1970; Pedersen 1971; Rasch et al 1971; Bachmann et al 1972; James 1972; Hinegardner 1972; Rees, Jones 1972; Macgregor, Walker 1973; Conger, Clinton 1973; Rasch, Balsano 1974; Rasch 1974; Mizuno, Macgregor 1974; Olmo, Morescalchi 1975; Shapiro 1976; Thiebaud, Fischberg 1977; Olmo, Morescalchi 1978; Horner, Macgregor 1983). These included Poecilia formosa (Amazon molly fish, 1.6 pg DNA/ cell), Gallus domestica (hen, 2.5 pg DNA/cell), Cyprinus carpio (carp, 3.4 pg DNA/cell), Pseudemus scripta (box turtle, 5.0 pg DNA/cell), Salmo gairdineri irideus rainbow trout, 5.2 pg DNA/cell), Xenopus laevis (African clawed toad, 6.0-6.3 pg DNA/cell), CF1 mouse (7.0-7.1 pg DNA/cell), Rana esculenta (European frog 13.8 pg DNA/cell), Plethodon glutinosus (salamander, 64 pg DNA/cell) Ambystoma mexicanum (Mexican axolotl, 77 pg DNA/cell), Notophthalmus viridescens (woodland newt, 90 pg DNA/cell), and Amphiuma means (Congo eel, 192.8 pg DNA/cell).

To obtain whole blood for cell smears, fishes, toads and salamanders were first chilled in an icebath. Blood was then taken directly from the heart with a heparinized capillary pipette. Thin, feather-edged blood films were prepared in the usual manner. Chicken blood from adult white leghorn hens was drawn into a heparinized syringe from a large wing vein and used to make 1-1.5 cm long cell smears near the end of a clean microscope slide. These blood films were quickly air-dried and stored in the dark at room temperature to serve as control preparations when needed for the addition of "test" blood smears or tissue imprints on the other end of the slide.

Each set of preparations to be stained by the Feulgen reaction was first fixed for 30 min in MFA (methanol, formaldehyde, glacial acetic acid, 85:10:5, v/v), followed by 15 min of washing in running tap water and three 5-min rinses in distilled water prior to their simultaneous hydrolysis in 5N HCl at 23 C for 30 min. After a 15-20 sec rinse in 0.01N HCl to minimize carry-over of strong acid, all slides were stained together for two hours in 1% Schiff leucofuchsin sulfurous acid reagent (pH 2.0-2.3), prepared as described elsewhere (Rasch et al 1971). After three 5-min rinses in freshly prepared sulfite water and washing for ten min in running tap water, slides were given three 5-min rinses in distilled water, dehydrated through a graded ethanol series, air-dried from absolute ethanol and stored in darkness for up to six months with no detectable loss of staining intensity, as determined by monthly measurements of a slide carrying both chicken and trout blood cells.

To minimize non-specific light loss due to scatter, all preparations were mounted just prior to measuring in matching refractive index liquids (n_D 1.524-1.532; RP Cargille Laboratories, Inc., Cedar Grove NJ 07009) for high resolution cytophotometry with a Vickers M86 scanning and integrating microdensitometer that has a 400-700 nm photomultiplier tube and uses an electronic masking system to allow operator selection of slide areas for measurement (Vickers Instruments, Inc., Malden MA 02148). As dictated by the size differences in the nuclei to be measured, either a x20/0.5 objective or a x100/1.25 oil immersion, achromatic objective was used with immersion oil (n_D 1.524) on the top lens of the condensor, the iris diaphragm of which was adjusted to match the N.A. of each objective by using the light meter of the instrument to determine the cut-off point beyond which further opening of the iris gave no further increase in transmittance. Although all of the integrated absorbance values for individual nuclei were obtained in arbitrary units (A.U.) within a 4 second interval from 50,000 point absorbance measurements performed within, or corrected to, the high resolution 1x1 scanning grid of the Vickers instrument, all three scanning patterns were used to accommodate the enormous range in sizes of the different types of nuclei that were measured with an 0.4 um scanning spot. Coefficients of variation of 0.6-0.7% were routinely obtained in 50 repeated scans of the same chicken blood cell nucleus and coefficients of variation of 1.8-2.5% regularly occur for sets of measurements of 50 different chicken or trout erythrocyte nuclei.

Because many large and intensely stained amphibian nuclei were too dense for scanning at 560 nm, near the absorption peak of the Feulgen-DNA dye complex, point absorption measurements were taken on each preparation by positioning the scanning spot in the center of a nucleus and selecting an appropriate wavelength to yield average absorbance values of 0.3 to 0.7 (Allison et al 1981). Measurements made at off-peak wavelengths or with different combinations of objectives and scanning patterns were corrected by measuring 25 Xenopus nuclei at 560 nm and again off-peak at either 600 nm or 620 nm. Integrated absorbances in A.U. were further corrected by the use of constants for k_d and k_a (Goldstein 1970; Goldstein 1971). Corrections for a glare error of 3% at 560 $\pm$ 17 nm and 2% at 620 $\pm$ 23 nm were made electronically as recommended by Bedi and Goldstein (1976). When the effect of spot size was evaluated (Goldstein 1971), the apparent integrated absorbances of toad and axolotl nuclei did not vary with the three smallest of the available scanning spots and thus, no further corrections were made for spot size or residual distributional error. All nuclear absorbance values were obtained by the subtractive method (Bedi and Goldstein 1976; Rasch, Rasch 1979; Miksche et al 1979).

Rapid data storage and subsequent retrieval of sets of measurements for transformation, statistical processing and display of regression curves were achieved with a Sol IIIA microcomputer that was directly coupled to the Vickers M86 with a custom-designed interface (Microproducts and Systems, Inc., Kingsport TN 37664), as described elsewhere (Rasch, Rasch 1979).

Results and Observations

Representative Feulgen-stained nuclei are shown in Figs. 2-5 for erythrocytes from several of the organisms used here as reference standards to derive calibration curves to validate our estimate of DNA content for the Drosophila pseudonurse cell chromosomes shown in Fig. 1.

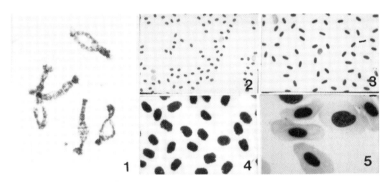

Figs. 1-5. Feulgen-stained chromosomes of Drosophila pseudonurse cell polytene chromosomes (Fig. 1) and blood cell nuclei of chicken (Fig. 2), trout (Fig. 3), axolotl (Fig. 4) and Congo eel (Fig. 5). All at 270X.

The validity for presuming an equivalency, or at least a constant proportionality between measured amounts of the Feulgen-DNA dye complex and absolute DNA content per cell can be addressed empirically by comparing the amounts of Feulgen staining found for "known" reference standards that have been selected because there is good agreement in the literature on the DNA content of these cells from direct biochemical determinations (Ris, Mirsky 1949). For example, a ratio of 2.52:1 is expected from a comparison of the DNA amounts in erythrocytes of Xenopus laevis (6.3 pg DNA; Thiebaud, Fischberg 1977) and chicken blood cell nuclei (2.5 pg DNA; Rasch et al 1971; Pogany et al 1981). Over a period of eight years, we have found a ratio of 2.56 ± 0.016:1 from 35 independent series of measurements of more than 3,625 individual nuclei from Xenopus and chicken erythrocytes. The observed ratio is within 1% of the expected ratio.

Similarly, when using blood cell nuclei from rainbow trout (5.2 pg DNA; Vendrely, Vendrely 1956; Rasch et al 1970; Hinegardner 1972) and chicken erythrocytes as internal reference standards, a ratio of 2.08:1 is expected. During the past five years, an observed ratio of 2.07 ± 0.015:1 was found from 17 separate series of measurements involving more than 1,620 individual nuclei. Again, the observed ratio agrees within 1% of the anticipated ratio. Interestingly, we estimated a 2C genome size of 5.2 pg DNA for rainbow trout some 20 years ago, using two wavelength cytophotometry (Ornstein 1952) with a custom designed Leitz microspectrophotometer that is described elsewhere (Rasch et al 1965).

Evidence in support of the accuracy of these relationships in the DNA content of trout and chicken nuclei comes from an observed ratio of 2.1 ± 0.01:1 (Lee GM, personal communication) that was obtained by flow cytofluorometric analysis of more than 60 different samples containing mixtures of isolated trout and chicken nuclei stained with DAPI (4'-6-diamidino-2-phenylindole), a DNA-specific fluorochrome (Coleman et al 1981; Lee et al 1984). Chicken and fish erythrocyte nuclei are now being included as internal reference standards by an increasing number of workers in the field of flow microfluorometry (Coulson et al 1977; Coulson, Tyndall 1978; Tannenbaum 1978; Vindelov et al 1980; Vindelov et al 1983; Lovett et al 1984).

Direct comparisons of flow microfluorometric DNA determinations with propidium iodide or DAPI and absorption cytophotometric measurements of the DNA content of Feulgen-stained mouse thymocytes and chick erythrocytes have demonstrated that these two quite different technologies yield essentially similar estimates of DNA content (Allison et al 1981). Furthermore, the use of at least two internal standards with Feulgen-stained preparations can provide absorption cytophotometric DNA values for mammalian cells with a precision equivalent to that claimed for flow cytometry with DNA-specific fluorochromes (Vindelov et al 1983; Lee et al 1984). In both cases, coefficients of variation of 1.5-3.0% can be achieved for non-dividing cell populations (Allison et al 1981).

A representative calibration curve is shown in Fig. 6 for reference standards that included sperm of Drosophila melanogaster (0.18 pg DNA), erythrocyte nuclei of the Amazon molly fish (1.6 pg DNA), chicken (2.5 pg DNA), carp (3.4 pg DNA), rainbow trout (5.2 pg DNA), Xenopus laevis (6.3 pg DNA) and human leucocytes (7.2 pg DNA). The data shown represent 195 replicated measurements at 560 nm of individual nuclei and yielded a linear regression from 0.18 pg DNA to 7.2 pg DNA with a zero intercept and a correlation coefficient of 0.996 between observed and expected values.

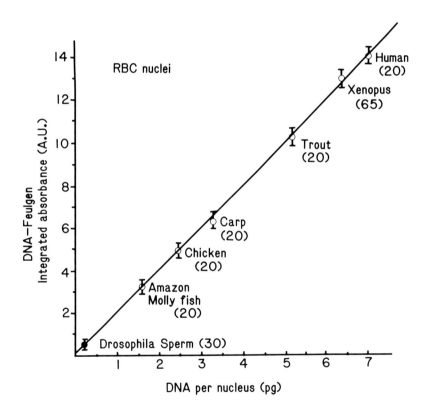

Fig. 6. Calibration curve of genome size estimates by DNA-Feulgen cytophotometry for blood cell nuclei of six vertebrates and Drosophila sperm. Values on the ordinate are expressed in arbitrary units (A.U.) of integrated absorbance at 560 nm. The standard deviation is shown by the vertical bars for each mean. Number of nuclei shown in parentheses.

The values shown in Fig. 7 demonstrate that there is a remarkably good, linear relationship between integrated absorbance values for individual, Feulgen-stained, fish or amphibian blood cell nuclei and biochemical estimates of the amount of DNA per cell for these animals, spanning a range from about 5 pg (rainbow trout) to almost 200 pg DNA (Congo eel). Again, the curve passes through zero and has a correlation coefficient of 0.993 between the observed and the expected values.

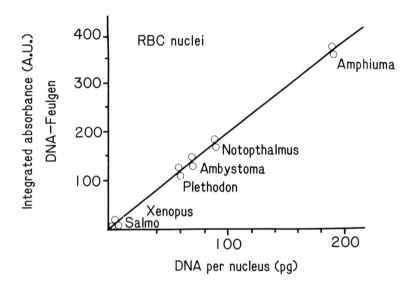

Fig. 7. Calibration curve of genome size estimates by DNA-Feulgen cytophotometry for five species of amphibians and rainbow trout. Values on the ordinate are expressed in arbitrary units (A.U.) of integrated absorbance at 620 nm. Each of the points shown represents the mean obtained from scans of 25-30 individual nuclei. Coefficients of variation ranged from 2.6% to 7.5%.

Using chicken and trout blood cell nuclei or rat kidney nuclei as reference standards, we have obtained similar data when estimating diploid genome sizes for several plant species ranging from 0.2 pg DNA for the green, coenobial alga Pediastrum boryanum (Millington, Rasch 1980) to 5 pg DNA per 2C nucleus for corn (Zea mays); 4 to 5 pg for cotton (Gossypium hirsutum); 18 pg DNA for horse bean (Vicia faba, Rasch et al 1959); 40 pg DNA for onion (Allium cepa, Rasch, Woodard 1959); 52-54 pg DNA for the spiderwort (Tradescantia paludosa); 60 to 61 pg DNA for Lilium henyri or Lilium speciosum; 88 pg DNA for the Lilium cultivar "Enchantment" (Rasch cited in Friedman et al 1982) and roughly 100 pg for the Easter lily, Lilium longiflorum (Rasch, Woodard 1959). The DNA estimates obtained in this way show good agreement with values cited elsewhere in the literature for these same plant species (McLeish, Sunderland 1961; Shapiro 1970; Rees, Jones 1972; Price et al 1973; Bennett, Smith 1976).

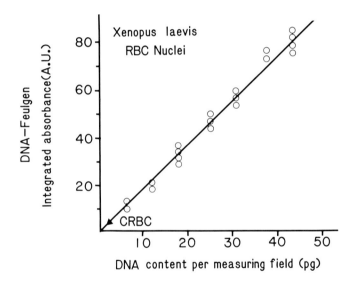

Fig. 8. Relationship between integrated absorbance at 560 nm and actual DNA content computed for increasing numbers of Xenopus laevis erythrocyte nuclei per measuring field.

By simply varying the number of nuclei that have a "known" DNA content within a given measuring field, we have been able to test the linearity in response of the Vickers photomultiplier tube, expressed as integrated absorbances in machine units to incremental differences in the amounts of DNA-Feulgen dye complex per scanning field. Fig. 8 shows the data from 45 sets of measurements at 560 nm of Xenopus blood cell nuclei, referenced to the mean Feulgen dye content determined for 50 individual chicken erythrocyte nuclei on the same slide. For the range tested here, 2.5 to 44.5 pg DNA, a correlation coefficient of 0.997 was obtained between observed values of Feulgen-DNA integrated absorbance and projected actual DNA content per measuring field.

The data displayed in Fig.9 for 74 separate sets of measurements at 560 nm of trout erythrocyte nuclei also were referenced to the mean Feulgen-DNA content of 50 individual chicken blood cell nuclei and yielded a correlation coefficient of 0.992 between observed and expected values in the range of 2.5 to 78 pg DNA per measuring field.

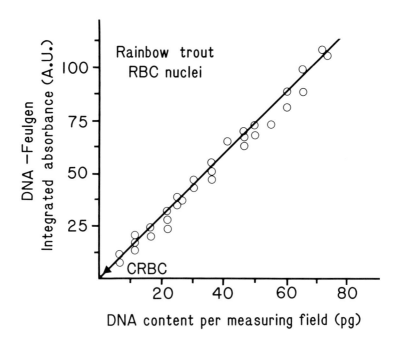

Fig. 9. Relationship between observed integrated absorbance at 560 nm and actual DNA content computed for increasing numbers of trout red blood cell nuclei per measuring field, referenced to the mean determined for 50 chicken blood cell nuclei on the same slide.

To extend the range of DNA values further, 24 sets of different numbers of nuclei from the terrestrial salamander Plethodon glutinosus were measured at 620 nm, spanning a range from 64 to 320 pg DNA per scanning field (Fig. 10). A correlation coefficent of 0.978 was found between observed values of integrated absorbance of Feulgen dye and projected actual DNA content per measuring field.

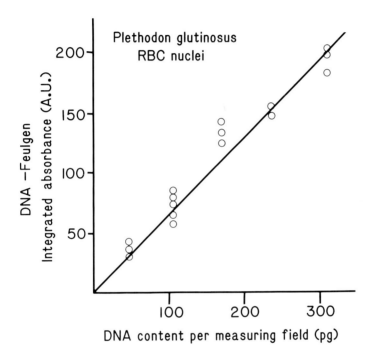

Fig. 10. Relationship between observed integrated absorbance at 620 nm and actual DNA content computed for increasing numbers of Plethodon blood cell nuclei per field.

Fig. 11 shows the relationship observed between estimates of DNA content from integrated absorbances at 620 nm and differences in the number of packets of Feulgen stain per measuring field using blood cell nuclei of the newt, Notophthalmus viridescens. A correlation coefficient of 0.989 was found for 44 separate sets of measurements that covered the range from 90 to 450 pg DNA per scanning field.

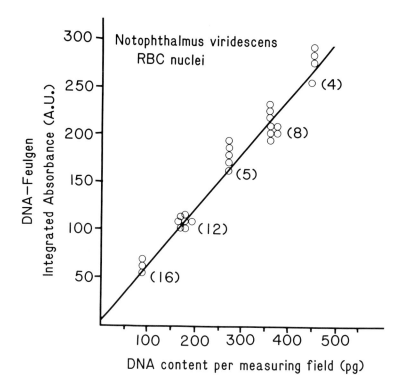

Fig. 11. Relationship between observed integrated absorbance at 620 nm and the actual DNA content computed for increasing numbers of Notophthalmus erythrocyte nuclei per measuring field.

Finally, as shown in Fig. 12, using different numbers of erythrocyte nuclei of Amphiuma in the scanning field, 70 separate sets of measurements made at 620nm with a x20 objective yielded a correlation coefficient of 0.999 between observed values for integrated absorbance of the Feulgen-DNA dye complex and projected actual DNA levels in the range from 193 to 2,120 pg DNA.

For each of the sets of measurements shown in Figs. 10, 11 and 12, point absorbance determinations were made at 620 nm in the manual operating mode of the Vickers instrument to ascertain that the density of individual nuclei did not exceed 0.7 (Allison et al 1981).

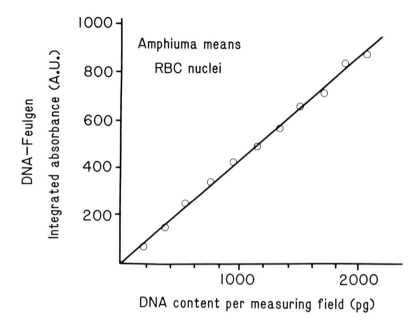

Fig. 12. Relationship between observed integrated absorbance expressed in arbitrary units (A.U.) and actual DNA content computed for increasing numbers of Amphiuma blood cell nuclei per measuring field.

Caveats and Applications

The calibration curves shown in Figs. 6 to 12 demonstrate an encouraging correlation between cytophotometric determination of DNA content based on Feulgen staining and values obtained for the same kinds of cells by direct chemical analysis (Mirsky, Ris 1951; Dawiid 1965; Shapiro 1970; Hinegardner, Rosen 1972; Rasch 1974; Shapiro 1976; Pogany et al 1981). By the appropriate use of a variety of cell types ranging from Drosophila sperm to blood cell nuclei from different fishes and amphibians and by varying the number of such test objects within the field scanned by an integrating microdensitometer, we have found that actual DNA content can be estimated reliably over a span of five orders of magnitude, i.e., from less than 0.2 pg (2×10^{-13} g) DNA to more than 2 ng (2×10^{-9} g) of DNA.

It is important to emphasize that the types of nuclei used to obtain the linear regressions shown in Figs. 6 to 12 are not only convenient test objects that can readily be prepared as smears of single cells or isolated nuclei, but they also exhibit essentially equivalent levels of chromatin condensation (Figs. 2 to 5). Marked differences in the degree of chromatin condensation can significantly influence the amounts of Feulgen-DNA dye complex bound and thus lead to spurious observations of apparent differences in DNA content, the so-called chromosome "compaction" error initially described by Hale (1963) and subsequently studied by a number of other workers (Mayall, Mendelsohn 1967; Garcia 1968; Mayall 1969; Garcia 1970; Noeske 1971). While some of the discussion on this issue involved specious, circular reasoning based on expectations of strict adherence to the DNA constancy hypothesis (cf Noeske 1971), it is nonetheless true that the Feulgen reaction for DNA, like many popular, DNA-specific fluorochromes, may show perturbations in dye stoichiometry when direct comparisons are made of staining levels in nuclei such as mouse thymocytes vs hepatocytes (Cowden, Curtis 1981) or fish erythrocytes vs hepatocytes (Rasch, Balsano 1974).

Classes of nuclei that show great differences in the amount of non-histone proteins and nuclear volumes or in which there are marked differences in DNA base ratios (Leemann, Ruch 1982; Hauser-Urfer, Leemann, Ruch 1982) can be expected to show deviations from expected levels of dye binding. Indeed, apparent discrepancies between observed levels of dye binding and actual DNA content may prove to be very sensitive indicators of alterations in chromatin organization and transcriptional activity.

A practical example of the use of DNA-Feulgen calibration curves is shown by computing that the Drosophila polytene chromosome set shown in Fig. 1 contains approximately 860 pg DNA, or more than 100 times as much DNA as the human diploid genome! When this amount of DNA is viewed in terms of a female haploid genome size of 0.2 pg DNA (Mulligan and Rasch 1980), it is reasonable to conclude that these chromosomes have undergone 11 cycles of DNA synthesis and are approximately at the 4096C level of endoreduplication.

Yet another use for reference standards is shown in Table 1 which illustrates the type of data obtained from DNA-Feulgen measurements of 2C hemocytes from Drosophila virilis to determine somatic cell genome size. Table 2 lists haploid genome size estimates for several species of Drosophila, including D. virilis and the Hawaiian species, D. grimshawi, that has roughly 40% more DNA than D. melanogaster (Rasch, cited in Laird 1973; Rasch et al 1977; Rasch EM, Kambysellis MP 1980, unpublished).

Table 1

Genome size estimates for Drosophila virilis

Specimen	DNA-Feulgen per nucleus mean $\pm$ S.E.		Number of nuclei	DNA per hemocyte (pg)	Somatic cell genome size* (x 10^6 kb)
Dv-1	2.263	0.013	83	0.67	0.58
Dv-2	2.290	0.012	100	0.68	0.58
Hen RBC	8.420	0.020	100	2.50	2.28

*Computed by assuming 9.13 x 10^8 nucleotide pairs per pg DNA (Britten and Davidson, 1971)

Table 2

Genome size estimates for species of Diptera

Organism	Haploid Genome size DNA (pg)	Molecular wt. haploid genome (x 10^{10} daltons)
Drosophila melanogaster	0.18	10.8
Drosophila hydei	0.20	12.0
Drosophila grimshawi	0.25	15.1
Drosophila virilis	0.34	20.5

Three internal reference standards (hen, trout and Xenopus) were used to determine genome size for several species of parasitoid wasps (Hymenoptera) and two lepidopteran species, the wax moth, Galleria melonella and the silk worm, Bombyx mori (Table 3). Both species of Habrobracon have genomes about the size of Drosophila melanogaster. Bombyx and Galleria, on the other hand, have C values of about 0.5 pg DNA (Rasch 1974; Rasch et al 1977; Rasch EM, Kumaran K 1977, unpublished).

Table 3

Genome size estimates for species of Hymenoptera & Lepidoptera

Organism	Haploid Genome size DNA (pg)	Molecular wt. haploid genome (x 10^{10} daltons)
Hymenoptera		
Habrobracon juglandis	0.16	9.6
Habrobracon serinopae	0.16	9.6
Nasonia (Mormoniella) vitripennis	0.34	20.5
Lepidoptera		
Bombyx mori	0.52	31.3
Galleria melonella	0.50	30.1

Unlike the size of insect genomes shown in Tables 1, 2 and 3, orthopteran species have large genomes (John, Hewitt 1966; John 1983). DNA values for hemocyte nuclei from the cave cricket, Hadenoecus and the camel cricket Ceuthophilus, are given in Tables 4 and 5 and compared with the genome size for several other orthopteran species in Table 6. DNA values for the Australian grasshopper, Acrida conica and the house cricket, Acheta domesticus were computed from data of John and Hewitt (1966) and Lima-de-Faria et al (1973). The marked variation in genome size among just a few orthopteran species warrants further study. The coupling of modern methods of molecular biology with the more classic methods of cytophotometry should provide interesting insights on the significance of genome size in the evolutionary success of these cytogenetically complex insects (White1973; 1980).

Table 4

Genome size estimates for the cave cricket Hadenoecus subterraneus

Specimen	DNA-Feulgen per nucleus mean ± S.E.		Number of nuclei	DNA per hemocyte (pg)	Somatic cell genome size* (x 10^6 kb)
H-1	10.48	0.030	95	3.11	2.84
H-2	10.95	0.050	75	3.25	2.97
H-3	10.17	0.052	75	3.02	2.76
H-4	10.55	0.049	75	3.13	2.84
H-5	10.85	0.045	75	3.22	2.94
H-6	10.15	0.037	75	3.01	2.75
			470	av. 3.12	av. 2.85
Rainbow Trout RBC	17.11	0.041	50	5.08	4.64
Xenopus RBC	22.31	0.038	150	6.62	6.04
Chicken RBC	8.42	0.020	100	2.50	2.28

*Computed by assuming 9.13 x 10^8 nucleotide pairs per pg DNA (Britten and Davidson, 1971)

Table 5

Genome size estimates for the camel cricket Ceuthophilus stygius

Specimen	DNA-Feulgen per nucleus mean ± S.E.		Number of nuclei	DNA per hemocyte (pg)	Somatic cell genome size* (x 10^6 kb)
C-1	63.33	0.229	110	18.80	17.16
C-2	65.61	0.151	75	19.48	17.78
C-3	62.80	0.076	75	18.65	17.03
C-4	65.33	0.219	110	19.39	17.70
			370	av. 19.08	av. 17.42
Human leukocytes	24.82	0.046	100	7.36	6.72
Xenopus RBC	22.31	0.038	150	6.62	6.04
Chicken RBC	8.42	0.020	100	2.50	2.28

*Computed by assuming 9.13 x 10^8 nucleotide pairs per pg DNA (Britten and Davidson, 1971)

Table 6

Genome size estimates for species of Orthoptera

Organism	Haploid Genome size DNA (pg)	Molecular wt. haploid genome ($\times 10^{10}$ daltons)
Hadenoecus subterraneous	1.55	93.6
Acheta (Gryllus) domesticus	2.00	120.5
Locusta migratoria	6.35	382.5
Ceuthophilus stygius	9.50	572.2
Acrida conica	12.55	755.9

Feulgen-DNA cytophotometry has been very useful for estimating genome size in a number of animals where only small amounts of material are available for analysis. For example, C values of 0.23-0.25 pg DNA have been determined for two species of water fleas, Daphnia pulex and Daphnia magna (Rasch EM, Stanton D, Hebert P 1983, unpublished). These values will be useful when screening naturally occurring populations of diploid and triploid clones of parthenogenetic Daphnia.

Other practical information that can be gained from the use of internal reference standards is shown in Table 7, which lists genome size estimates for a number of species of birds, many of which are endangered, such as the California condor, the Puerto Rican parrot, the everglade kite and several species of cranes whose blood samples were obtained from Dr. George Archibald of the International Crane Foundation in Baraboo, Wisconsin and Dr. George Gee at the Patuxent Wildlife Research Center of the Fish and Wildlife Service of the U.S. Department of the Interior at Laurel, Maryland. As also noted by Bachmann and Harrington (1972), despite the extensive speciation among birds, genome sizes in this order are remarkably consistent. In some cases, the small difference in amounts of DNA between female (ZW) and male (ZZ) cranes can be detected by cytophotometry (Rasch 1976). Since cranes and many other species of endangered birds lack obvious sexual dimorphisms, quantitative cytophotometric procedures that can be applied to preparations from a single drop of blood may prove very useful for sex determinations on individual birds when establishing flocks for captive breeding.

Table 7

Genome size estimates for species of birds

Organism	Number of specimens	Haploid Genome size DNA (pg)	Molecular wt. haploid genome (x 10^{10} daltons)
Common black crow	5	1.25	75.3
California condor	1	1.51	91.0
Everglade kite	8	1.41	84.9
Hispanolan parrot	6	1.62	97.6
Puerto Rican parrot	4	1.58	95.2
European crane	4	1.51	91.0
Sandhill crane	6	1.54	92.8
Siberian crane	2	1.55	93.4
White naped crane	4	1.52	91.4
Whooping crane	6	1.59	95.8
Domestic chicken	6	1.25	75.3

As we have previously shown, genome size estimates made by Feulgen-DNA cytophotometry for *Drosophila melanogaster* (Rasch et al 1971; Mulligan, Rasch 1980) and for the silk worm, *Bombyx mori*, agree within a few percent of the estimates of haploid genome size for these organisms obtained by molecular biologists using DNA reassociation kinetics (Laird 1973; Gage 1974). Now, the size of the haploid genome for yet another insect species, the honey bee, Apis mellifera, has been determined by DNA-Feulgen cytophotometry of more than 100 individual sperm from each of two different drones. Using both chicken and trout blood cell nuclei as reference standards of 2.5 pg and 5.2 pg DNA, respectively, the haploid genome of the honey bee is 0.17 pg DNA, or roughly 1.02×10^{11} daltons (Rasch unpublished). This value agrees very well with the estimate of 1.14×10^{11} daltons that was computed to be the size of the honey bee haploid genome by Jordon and Brosemer (1974) from their analysis of reassociation kinetics for the DNA isolated from worker bees. It should be interesting to compare C-values among other major groups of Hymenoptera to see whether or not the great diversity in form and social behavior of these insects will be reflected in genome size.

In summary, for the past 35 years a small but significant number of biologists have considered the light microscope to be an instrument particularly well suited for the chemical analysis of cells and tissues, in addition to its routine use for purely morphological studies (Swift 1966; Grove 1979; Haselmann 1983; Chieco, Boor 1984). Despite a decline in the number of its practitioners, static absorption microspectrophotometry is especially useful when quantitative determinations of DNA levels are needed for cell systems in which small sample size and/or tissue heterogeneity preclude direct biochemical analysis or the use of flow cytometry. With the application of microcomputer technology to modern, solid-state instruments for absorption cytophotometry and the use of reference standards, as well as appropriate corrective techniques, this type of analytical microscopy should continue to make important contributions to the fields of genetics, developmental cytology, evolutionary biology and clinical pathology.

Acknowledgments

This research was supported in part by grants from the National Science Foundation (DEB 77-03257 and PCM 81-03250), the National Audubon Society and the Veterans Administration (1A 74 111 430100). Blood smears from Hispaniolan parrots, Puerto Rican parrots and everglade kites were kindly provided by Dr. Noel Snyder (U.S. Fish and Wildlife Service). Blood smears from the California condor were supplied by Dr. Arden Bercovitz, Research Department, San Diego Zoo. Drone bees were from the apiary of Ms. Sherry Apple (Memphis, TN).

References

Allison DC, Ridolpho PF, Rasch EM, Rasch RW, Johnson TS (1981). Increased accuracy of absorption cytophotometric DNA values by control of stain intensity. J. Histochem Cytochem 29: 1219.
Altman K (1973). Quantitation in histochemistry: A review of some commercially available microdensitometers. Histochem J. 7:375.
Atkin NB (1969). Perimodal variation of DNA values of normal and malignant cells. Acta Cytol 13:270.
Atkin NB (1976). Prognostic significance of ploidy level in human tumors 1. Carcinoma of the uterus. J Nat Cancer Inst. 56:909.
Atkin NB, Mattinson G, Becak W, Ohno S (1965). The comparative DNA content of 19 species of placental mammals, reptiles and birds. Chromosoma 17:1.
Bachmann K (1970). Feulgen slope determinations of urodele nuclear DNA amounts. Histochemie 22:289.
Bachmann K (1972). Genome size in mammals. Chromosoma 37:85.
Bachmann K, Harrington BA, Craig JB (1972). Genome size in birds. Chromosoma 37:405.
Bachmann K, Rheinsmith EL (1973). Nuclear DNA amounts in Pacific Crustacea. Chromosoma 43:225.
Bedi KS, Goldstein DJ (1976). Apparent anomalies in nuclear Feulgen-DNA contents. Role of systematic microdensitometric errors. J Cell Biol 71:68.
Bennett MD (1972). Nuclear DNA content and minimum generation time in herbaceous plants. Proc Roy Soc Lond B 181: 109.
Bennett MD, Smith JB (1976). Nuclear DNA amounts in angiosperms. Phil Trans Roy Soc Lond B 274:227.
Britten RJ, Davidson EH (1971). Repetitive and non-repetitive DNA sequences and a speculation on the origins of evolutionary novelty. Quart Rev Biol 46:111.
Cavanagh JB (1979). DNA polyploidy and mitotic spindle abnormalities in astrocytes. In Pattison JR, Bitensky L, Chayen J (eds): "Quantitative Cytochemistry and its Applications", London: Academic Press, p 23.
Chieco P, Boor PJ (1984). Quantitative histochemistry in pathology and toxicology. An evaluation of the original two-wavelength method of Ornstein. Lab Invest 50:355.
Clark G, Kasten FH (1983). "History of Staining", 3rd ed, Baltimore: Williams and Wilkins, p 240.

Coleman AW, Maguire MJ, Coleman JR (1981). Mithramycin and 4'-6-diamidino-2-phenylindole (DAPI)-DNA staining for fluorescence microspectrophotometric measurement of DNA in nuclei, plastids and virus particles. J Histochem Cytochem 29: 959.
Conger AD, Clinton JH (1973). Nuclear volumes, DNA contents and radiosensitivity in whole-body irradiated amphibians. Radiation Res 54:69.
Coulson PB, Bishop AO, Lenarduzzi R (1977). Quantitation of cellular deoxyribonucleic acid by flow microfluorometry. J Histochem Cytochem 25:1147.
Coulson PB, Tyndall R (1978). Quantitation by flow microfluorometry of total cellular DNA in Ancanthamoeba. J Histochem Cytochem 26: 713.
Cowden RR, Curtis SK (1981). Microfluorometric investigations of chromatin structure I. Evaluation of nine DNA-specific fluorochromes as probes of chromatin organization. Histochemistry 72:11.
Cullis CA (1973). DNA amounts in the nuclei of Paramecium bursaria. Chromosoma 40:127.
Dawid I (1965). Deoxyribonucleic acid in amphibian eggs. J Molec Biol 12:581.
Dhillon SS, Berlyn GP, Miksche JP (1977). Requirement of an internal standard for microspectrophotometric measurement of DNA. Amer J Bot 64:117.
Duijndam WAL, vanDuijn P (1975). The influence of chromatin compaction on the stoichiometry of the Feulgen-Schiff procedure studied in model films II. Investigations on films containing condensed or swollen chicken erythrocyte nuclei. J Histochem Cytochem 23:891.
Duijndam WAL, Smeulders AWM, vanDuijn P, Verweij AC (1980a). Optical errors in scanning stage cytophotometry I. Procedures for correcting apparent integrated absorbance values for distributional glare and diffraction errors. J Histochem Cytochem 28:388.
Duijndam WAL, vanDuijn P, Riddersma SH (1980b). Optical errors in scanning stage cytophotometry II. Applications of correction factor for residual distributional error, glare and diffraction error in practical cytophotometry. J Histochem Cytochem 28:395.
Friedman BE, Bouchard RA, Stern H (1982). DNA sequences repaired at pachytene exhibit strong homology among distantly related higher plants. Chromosoma 87:409.
Gage LP (1974). The Bombyx mori genome: analysis by DNA reassociation kinetics. Chromosoma 45:27.

Gall JG (1981). Chromosome structure and the C-value paradox. J Cell Biol 91:3s.
Garcia AM (1970). Stoichiometry of dye binding versus degree of chromatin coiling. In Weid GL, Bahr GF (eds): "Introduction to Quantitative Cytochemistry", vol 2, New York: Academic Press Inc, p 153.
Goldstein DJ (1970). Aspects of scanning microdensitometry I. Stray light (glare). J Microsc 92:1.
Goldstein DJ (1971). Aspects of scanning microdensitometry II. Spot size, focus, and illumination. J. Microsc 93:15.
Grove GL (1979). Applications of microspectrophotometry to biomedical research. J Histochem Cytochem 27:1375.
Hale AJ (1963). The leucocyte as a possible exception to the theory of deoxyribonucleic acid constancy. J Path Bact 85:311.
Hale AJ (1966). Feulgen microspectrophotometry and its correlation with other cytochemical methods. In Weid GL (ed): "Introduction to Quantitative Cytochemistry", New York: Academic Press, p 183.
Haselmann H (1983). The future of light microscopy. Proc. Roy Microsc Soc 18:229.
Hauser-Urfer I, Leemann U, Ruch F (1982). Cytofluorometric determination of the DNA base content in human chromosomes with quinacrine mustard, Hoechst 33258, DAPI and mithramycin. Exp Cell Res 142:455.
Hinegardner R, Rosen DE (1972). Cellular DNA content and the evolution of teleostean fishes. Amer Nat 106:621.
Horner HA, Macgregor HC (1983). C value and cell volume: their significance in the evolution and development of amphibians. J Cell Sci 63:135.
James J (1965). Constancy of nuclear DNA and accuracy of cytophotometric measurements. Cytogenetics 4:19.
James J (1972). DNA constancy and chromatin structure in some cell nuclei of Amphiuma. Histochem J 4:181.
John B (1983). The role of chromosome change in the evolution of orthopteroid insects. In Sharma AK, Sharma A (eds): "Chromosomes in Evolution of Eukaryotic Groups", Boca Raton, FL: CRC Press Inc, p 1.
John B, Hewitt GM (1966). Karyotype stability and DNA variability in the Acridae. Chromosoma 20:155.
Jordon RA, Brosemer RW (1974). Characterization of DNA from three bee species. J Insect Physiol 20:2513.
Kasten FH (1960) The chemistry of Schiff's reagent. Intern Rev Cytol 10:1.
Kavenoff R, Zimm BH (1973). Chromosome sized DNA molecules from Drosophila. Chromosoma 41:1.

King RC, Riley SF, Cassidy JD, White PE, Pai YK (1981). Giant polytene chromosomes from ovaries of a Drosophila mutant. Science 212:441.
Kjellstrand P (1980). Mechanisms of the Feulgen hydrolysis. J Microsc 119:391.
Laird CD (1971). Chromatid structure: Relationship between DNA content and nucleotide sequence diversity. Chromosoma 32: 378.
Laird CD (1973). DNA of Drosophila chromosomes. Ann Rev Genet 7:177.
Lee GM, Thornthwaite JT, Rasch EM (1984). Picogram per cell determination of DNA by flow cytometry. Anal Biochem 137:221.
Leemann U, Ruch F (1982). Cytofluorometric determination of DNA base content in plant nuclei and chromosomes by the fluorochromes DAPI and chromomycin A3. Exp Cell Res 140:275
Lima-de-Faria A, Gustafsson AT, Jaworska H (1973). Amplification of ribosomal DNA in Acheta II. The number of nucleotide pairs of the chromosomes and chromomeres involved in amplification. Hereditas 73:119.
Lovett EJ, Schnitzer B, Keren DF, Flint A, Hudson JL, McClathchey KD (1984). Application of flow cytometry to diagnostic pathology. Lab Invest 50:115.
Macgregor HC, Walker MH (1973). The arrangement of chromosomes in nuclei of sperm from plethodontid salamanders. Chromosoma 40:243.
Mayall BH (1969). Deoxyribonucleic acid cytophotometry of stained human leukocytes I. Differences among cell types. J Histochem Cytochem 17:249.
Mayall BH, Mendelsohn ML (1970). Errors in absorption cytophotometry. Some theoretical and practical considerations. In Weid GL, Bahr GF (eds): "Introduction to Quantitative Cytochemistry", vol 2, New York: Academic Press Inc, p 171.
McLeish J, Sunderland N (1961). Measurements of deoxyribonucleic acid (DNA) in higher plants by Feulgen photometry and chemical methods. Exp Cell Res 24:527.
Miksche JP (1968). Quantitative study of intraspecific variation of DNA per cell in Picea glauca and Pinus banksiana. Can J Genet Cytol 10:590.
Miksche JP, Dhillon SS, Berlyn GP, Landauer KC (1979). Nonspecific light loss and intrinsic DNA variation problems associated with Feulgen DNA cytophotometry. J Histochem Cytochem 27: 1377.

Millington WF, Rasch EM (1980). Microspectrophotometric analysis of mitosis and DNA synthesis associated with colony formation in Pediastrum boryanum (Chlorophyceae). J Phycol 16:177.

Mirsky AE, Ris H (1951). Desoxyribonucleic acid content of animals cells and its evolutionary significance. J Gen Physiol 34:451.

Mizuno S, Macgregor HC (1974). Chromosomes, DNA sequences and evolution in salamanders of the genus Plethodon. Chromosoma 48:239.

Mulligan PK, Rasch EM (1980). The determination of genome size in male and female germ cells of Drosophila melanogaster by DNA-Feulgen cytophotometry. Histochemistry 66:11.

Noeske K (1971). Discrepancies between cytophotometric Feulgen values and deoxyribonucleic acid content. J Histochem Cytochem 19:169.

Olmo E, Morescalchi A (1975). Evolution of the genome and cell size in salamanders. Experientia 31:804.

Olmo E, Morescalchi A (1978). Genome and cell sizes in frogs: a comparison with salamanders. Experientia 34:44.

Ohno S, Atkin NB (1966). Comparative DNA values and chromosome complements of eight species of fishes. Chromosoma 18:455.

Ornstein L (1952). Absorption microphotometry of irregular-shaped objects. Lab Invest 1:250.

Pedersen RA (1971). DNA content, ribosomal gene multiplicity and cell size in fish. J Exp Zool 177:65.

Piller H (1977). "Microscope Photometry", Berlin: Springer Verlag, p.179.

Pogany GC, Corzett M, Weston S, Balhorn R (1981). DNA and protein content of mouse sperm. Implications regarding sperm chromatin structure. Exp Cell Res 136:127.

Pollister AW (1952). Microspectrophotometry of fixed cells by visible light. Lab Invest. 1:231.

Pollister AW, Swift H (1950). Molecular orientation and intracellular photometric analysis. Science 111:68.

Price HJ, Sparrow AH, Nauman AF (1973). Evolutionary and developmental considerations of the variability of nuclear parameters in higher plants I. Genome volume, interphase chromosome volume and estimated DNA content of 236 gymnosperms. Brookhaven Symp Biol 25:390.

Prescott DM (1983). The C-value paradox and genes in ciliated protozoa. In McIntosh JR (ed): "Modern Cell Biology" vol 2, New York: Alan R. Liss, p 329.

Rasch EM (1968). Use of erythrocyte deoxyribonucleic acid-Feulgen levels as an index of triploidy in naturally-occurring interspecific hybrids of poeciliid fishes. J. Histochem Cytochem 16:508.

Rasch EM, Barr HJ (1969). DNA-Feulgen cytophotometry of Drosophila hemocyte nuclei. J. Histochem Cytochem 17:187.

Rasch EM (1970). DNA cytophotometry of salivary gland nuclei and other tissue systems in dipteran larvae. In Weid GL, Bahr GF (eds): "Introduction to Quantitative Cytochemistry", vol 2, New York: Academic Press, p 357.

Rasch EM (1974). The DNA content of sperm and hemocyte nuclei of the silkworm Bombyx mori. Chromosoma 45:1.

Rasch EM (1976). Use of deoxyribonucleic acid-Feulgen cytophotometry for sex identification in juvenile cranes (Aves: Gruiformes). J Histochem Cytochem 24: 607.

Rasch EM, Balsano JS (1974). Biochemical and cytogenetic studies of Poecilia from eastern Mexico II. Frequency, perpetuation and probable origin of triploid genomes in females associated with Poecilia formosa, Rev Biol Trop 21:351.

Rasch EM, Barr HJ, Rasch RW (1971). The DNA content of sperm of Drosophila melanogaster. Chromosoma 33:1.

Rasch EM, Cassidy JD, King RC (1977). Evidence for dosage compensation in parthenogenetic Hymenoptera. Chromosoma 59:232.

Rasch EM, Darnell RM, Kallman KD, Abramoff P (1965). Cytophotometric evidence for triploidy in hybrids of the gynogenetic fish, Poecilia formosa. J Exp Zool 169:155.

Rasch EM, King RC, Rasch RW (1984). Cytophotometric studies on cells from the ovaries of otu mutants of Drosophila melanogaster. Histochemistry (in press).

Rasch EM, Monaco PJ, Balsano JS (1982). Cytophotometric and autoradiographic evidence for functional apomixis in a gynogenetic fish, Poecilia formosa and its related, triploid unisexuals. Histochemistry 73:515.

Rasch EM, Prehn LM, Rasch RW (1970). Cytogenetic studies of Poecilia (Pisces) II. Triploidy and DNA levels in naturally occurring populations associated with the gynogenetic teleost, Poecilia formosa (Girard). Chromosoma 31:18.

Rasch EM, Rasch RW (1979). Applications of microcomputer technology to cytophotometry. J. Histochem Cytochem 27:1384.

Rasch EM, Rasch RW (1981). Cytophotometric determination of genome size for two species of cave crickets (Orthoptera, Rhaphidophoridae). J Histochem Cytochem 29:885.

Rasch EM, Swift H, Klein RM (1959). Nucleoprotein changes in plant tumor growth. J Biophys Biochem Cytol 6:11.
Rasch EM, Woodard JW (1959). Basic proteins of plant nuclei during normals and pathological cell growth. J Biophys Biochem Cytol 6:263.
Rasch RW, Rasch EM (1973). Kinetics of hydrolysis during the Feulgen reaction for DNA. A Reevaluation. J. Histochem Cytochem 21:1053.
Rees H, Jones RN (1972). The origin of the wide species variation in nuclear DNA amount. Intern Rev Cytol 32:53.
Rees H, Shaw DD, Wilkinson (1978). Nuclear DNA variation among acridid grasshoppers. Proc Roy Soc London B 202:517.
Ris H, Mirsky AE (1949). Quantitative cytochemical determination of desoxyribonucleic acid with the Feulgen nucleal reaction. J Gen Physiol 33:125.
Sandritter W (1979). A review of nucleic acid cytophotometry in general pathology. In Pattison JR, Bitensky L, Chayen J (eds): "Quantitative Cytochemistry and Its Applications", London: Academic Press, p 1.
Shapiro HS (1970). Deoxyribonucleic acid content per cell of various organisms. In Sober HA, Harte R (eds): "Handbook of Biochemistry", 2nd ed. Boca Raton, FL: Chemical Rubber Company, p H104.
Shapiro HS (1976). DNA content of chordate cell nuclei. In Altman PL, Katz DD (eds): "Cell Biology", Bethesda: Fed Amer Soc Exp Biol, p 367.
Sklarew RJ (1982). Simultaneous Feulgen densitometry and autoradiographic grain counting with the Quantimet 720A image-analysis system I. Estimation of nuclear DNA content in ^{3}H-thymidine-labeled cells. J Histochem Cytochem 30:35.
Sklarew RJ (1983). Simultaneous Feulgen densitometry and autoradiographic grain counting with the Quantimet 720D image-analysis system III. Improvements in Feulgen densitometry. J Histochem Cytochem 31:1224.
Swift H (1950a). The desoxyribose nucleic acid of animal nuclei. Physiol Zool 23:169.
Swift H (1950b). The constancy of desoxyribose nucleic acid in plant nuclei. Proc Nat Acad Sci (Wash) 36:642.
Swift H (1953). Quantitative aspects of nuclear nucleoproteins. Intern Rev Cytol 2:1.
Swift H (1966). Analytical microscopy of biological materials. In Weid GL (ed): "Introduction to Quantitative Cytochemistry", New York: Academic Press, p 1.
Swift H, Kleinfeld R (1953). DNA in grasshopper spermatogenesis, oogenesis and cleavages. Physiol Zool 16:301.

Swift H, Rasch EM, Pollister AW (1969). Microphotometry with visible light. In Pollister AW (ed): "Physical Techniques in Biological Research", vol 3 part C "Cells and Tissues", New York: Academic Press, p 201.

Tannenbaum E, Cassidy M, Alabaster O, Herman C (1978). Measurement of cellular DNA mass by flow microfluorometry with the use of a biological internal standard. J Histochem Cytochem 26:145.

Thiebaud C, Fischberg M (1977). DNA content in the genus Xenopus. Chromosoma 59:253.

Thomson KS, Gall JG, Coggins LW (1973). Nuclear DNA content of coelacanth erythrocytes. Nature 241:126.

Vaughn JC, Locy RD (1969). Changing nuclear histone patterns during development III. The deoxyribonucleic acid content of spermatogenic cells in the crab Emerita analoga. J Histochem Cytochem 17:591.

Vendrely R, Vendrely C (1956). The results of cytophotometry in the study of deoxyribonucleic acid (DNA) content of the nucleus. Intern Rev Cytol 5:171.

Vindelov LL, Christensen IJ, Jensen G (1980). High resolution ploidy determination by flow cytometric DNA analysis with two internal standards. Basic Appl Histochem 24:271.

Vindelov LL, Christensen IJ, Nissen NI (1983). Standardization of high resolution flow cytometric DNA analysis by the simultaneous use of chicken and trout red blood cells as internal reference standards. Cytometry 3:328.

White MJD (1973). "Animal Cytology and Evolution", London: Cambridge University Press, p 420.

White MJD (1980). Meiotic mechanisms in a parthenogenetic grasshopper species and its hybrids with related bisexual species. Genetica 52/53:379.

REFINEMENTS IN ABSORPTION-CYTOMETRIC MEASUREMENTS OF
CELLULAR DNA CONTENT

David C. Allison, M.D., Ph.D.

Department of Surgery
Albuquerque VAMC and the University of New Mexico
Albuquerque, New Mexico 87108

Absorption cytophotometric measurements of Feulgen stained cells have several advantages for determination of cellular DNA content in various types of experiments. Measurements can be performed on rare cells selected on the basis of morphologic criteria. It is possible to measure other parameters as well as DNA content, such as cell protein content (Tas et al, 1980) or trypan blue viability (Allison, Ridolpho 1980), on the same cells. Most importantly, autoradiographs of the Feulgen-stained cells can be prepared, allowing determination of both DNA content and isotopic labeling of the same cells (Allison et al, 1983; Cooper et al, 1968).

Several problems must be resolved, however, in order to take full advantage of the absorption-cytophotometric method of measuring cellular DNA content. These problems include heterogeneous distribution of stain within individual nuclei (Duijndam et al, 1980; Goldstein 1971; Mayall, Mendelsohn 1970); variations in stain intensity between adjacent nuclei because of differences in nuclear size or stain content (Allison, et al, 1981); differences in average Feulgen-stain intensity for all nuclei in one area of a slide, compared to the average stain intensity for all nuclei in another area of the same slide (Atkin 1969); optical errors such as glare and diffraction (Duijndam et al, 1980; Goldstein 1970; Mayall, Mendelsohn 1970); and the requirement for rapid data transformation, analysis, and storage. Fortunately, advances in instrumentation and in methods of performing absorption measurements now make it possible to measure cellular DNA content with this technique with accuracy and rapidity sufficient for most biological experiments. A brief review of this method and of approaches to the above problems follows.

Heterogeneous distribution of stain within a nucleus.
The heterogeneous distribution of Feulgen stain within a nucleus precludes making a single measurement of light transmission in order to obtain an estimate of total nuclear stain content. This is because Lambert-Beer's law states that the amount of substance detected in a photometric measurement is proportional to the absorbance (E), which is the negative logarithm of the transmission (T):

$$E = -\ln T = (K)(C)(1),$$

where K is the absorption coefficient, C is the concentration of the light absorbing substance, and l is the specimen thickness. Thus, a single transmission measurement of the total Feulgen stain content of a heterogeneously stained nucleus will yield an arithmetic sum of the transmissions in different areas of this nucleus. This arithmetic sum is not proportional to the true integrated absorbance of this nucleus, however, since the integrated absorbance is proportional to the sum of the negative logarithms of the transmissions present in each different area of this nucleus.

Early workers showed considerable insight into this problem of "distributional error." In 1950, Swift, using a "plug" method which entailed making separate absorption measurements over multiple small areas of the same nucleus, showed constancy of Feulgen stain content in the nuclei of somatic plant cells. An extension of this approach led to the development of "flying-spot" cytometers. The Vickers M85 microdensitometer, which we use in our work, is an example of such a cytometer. This instrument makes point absorbance measurements with a small measuring spot of variable size (0.4 μm selected for use in these studies). In its scanning mode, the instrument automatically makes up to 50,000 point absorbance measurements in 5 seconds over a selected slide area that is defined by a mask. The sum of these point absorbances, the "integrated absorbance", is proportional to the true specimen absorbance. In practice, the Vickers M85 microdensitometer overcomes the problem of errors due to heterogeneous distribution of stain within limits that are acceptable for most biological experiments. Additional, and more complex, corrections for residual distributional error have been worked out for the Vickers M85 and for other types of absorption cytometers (Duijndam 1980; Goldstein 1971), and these can be used if needed.

Microcomputer transformation, storage, and analysis of data. In 1979, a direct interface of a microcomputer to a Vickers M85 cytometer was described (Rasch, Rasch 1979). This development, together with further software modifications (Allison et al, 1981), has greatly increased the utility of the Vickers M85 for multiple measurements of integrated absorbances that are often required for biological experiments.

Glare and diffraction errors. All absorption cytometers have a certain percentage of stray light, which increases the recorded transmission in an absorption measurement and leads to artifactually low absorbance values. In most modern cytometers, including the Vickers M85 instrument, the glare level can be reduced to approximately 3-4% by the use of "stops", proper objectives, and clean optics (Goldstein 1970). This % glare error, however, is a minimal estimate of the actual reduction of integrated-absorbance values. This small glare error can be achieved only for Feulgen-stained nuclei with very low average optical densities; the error increases exponentially as the optical density increases (Goldstein 1970; Mayall, Mendelsohn 1970). For example it has been estimated that the measured integrated absorbance of a Feulgen-stained nucleus with an average optical density of 1.1 with a 2% instrument glare level will only be approximately 90% of its true value (Goldstein 1970).

Diffraction at sharply stained interfaces, also leads to underestimation of the true integrated absorbance of a specimen. This is because some of the diffracted light reaches the recording photometer, artifactually increasing the transmission. This type of error is very important for absorption measurements performed on intensely stained small objects, as has been pointed out in two recent publications (Duijndam et al, 1980; Goldstein 1982). However, for measurements of integrated absorbances of relatively large objects, such as whole nuclei of mammalian or avian cells, both diffraction and glare errors can be reduced to acceptable limits if attention is paid to proper optics, reduction of stain darkness, and other simple corrections which are discussed below.

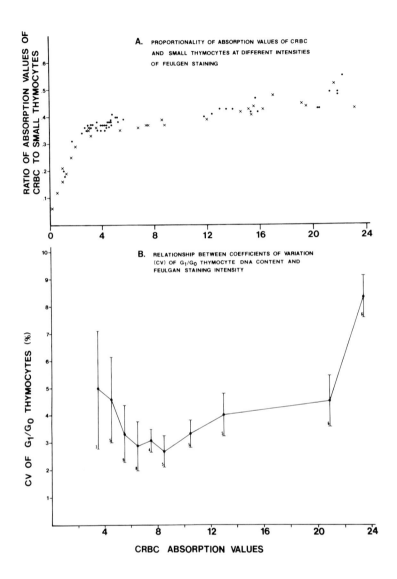

Figure 1. Integrated absorbances over a wide range of Feulgen stain intensities, measured at 560 nm. (A) integrated absorbance measurements of 10 small mouse thymocytes and 10 adjacent chicken erythrocytes on 70 slides Feulgen-stained to various intensities. (B) CVs of integrated absorbances of 40 G0/G1 thymocytes plotted against the average chicken erythrocyte integrated absorbance for 56 slides Feulgen-stained to various intensities. A linear relationship exists

between erythrocyte and thymocyte integrated absorbances (A), and the CVs of the G0/G1 thymocyte absorbances are minimal (B) on slides with chicken erythrocyte absorbances between 6 and 12 units. (Reprinted by permission of publisher from "Increased Accuracy of Absorption Cytophotometric DNA Values by Control of Stain Intensity", by Allison et al, J Histochem Cytochem 29:1219-1228, 1981, copyright by the Histochemical Society, Inc.)

Because glare and diffraction errors increase for darkly stained cells, we first decided to test whether decreasing the intensity of Feulgen staining would decrease errors in integrated-absorbance measurements (Allison et al, 1981). After nuclei on slides prepared from a mixture of chicken erythrocytes and mouse thymocytes were hydrolyzed for 1 hr in 4 N HCl at 20°C, we found that decreasing the incubation time in Schiff's reagent (Sigma, St. Louis MO) and/or lowering the pH (<1.4) of the Schiff's reagent decreased the intensity of Feulgen staining of these nuclei. We therefore Feulgen stained the nuclei of 70 slides under conditions selected to produce a wide variety of average stain intensities for each slide. We then measured the integrated absorbances at 560 nm (near the absorption maximum of the Feulgen stain) for 10 small mouse thymocytes and 10 adjacent chicken erythrocytes of each slide. The ratio of the chicken erythrocyte:mouse thymocyte integrated absorbances for each slide is plotted in Figure 1A against the average integrated absorbance of chicken erythrocytes (expressed in units taken directly from the cytophotometer) on the same slide.

On faintly stained slides, in which the erythrocyte absorbance was less than 4 units, the ratio of erythrocyte:thymocyte integrated absorbances fell to zero because the integrated-absorbance values obtained for the chicken erythrocytes were low relative to the integrated absorbances of the more intensely stained thymocytes (Fig. 1A). On lightly stained slides, in which the chicken erythrocyte integrated absorbances ranged between 4 and 11 units, the erythrocyte:mouse thymocyte integrated-absorbance ratio was stable at 0.37, a value close to the estimated ratio of the DNA content for the two species. On intensely stained slides with erythrocyte integrated absorbances greater than 20 units, the ratio for chicken erythrocytes:mouse thymocytes was high (0.47) and variable, presumably because of increased glare and diffraction errors in the measurements performed on the intensely stained small thymocytes.

To examine whether integrated-absorbance measurements of mouse thymocytes of all sizes could be improved by alterations in the stain intensity, the integrated absorbances of 40 G0/G1 mouse thymocytes of varying sizes and 24 adjacent chicken erythrocytes were measured on each of 56 slides Feulgen-stained to different stain intensities. In Figure 1B, the average coefficients of variation (CVs) of the thymocyte integrated absorbances for each slide are plotted against the average of the chicken erythrocyte integrated absorbances from the same slide. The CVs of the thymocyte integrated absorbances are at a minimum on slides that have chicken erythrocyte integrated-absorbance values between 6 and 14 units (Fig. 1B). The CVs of the thymocyte integrated absorbances increased for measurements performed on very faintly (chicken erythrocyte absorbance <4 units) or on intensely stained slides (chicken erythrocyte absorbance >12 units). Figures 1A and 1B clearly show that decreasing the Feulgen stain intensity can increase the accuracy of integrated-absorbance measurements, presumably because of the decrease in diffraction and glare errors.

We felt that it should be possible to obtain an improvement in the CVs of integrated-absorption measurements, similar to that achieved by decreases in the total stain content of Feulgen-stained nuclei (Fig. 1), if the integrated absorbance of intensely stained cells were to be measured at off-peak wavelengths of light (Swift 1950; Fand, Spencer 1968). If this is the case, the accuracy of integrated-absorbance measurements could be increased without cumbersome modifications of the staining procedure. To test this, slides of a mixture of mouse thymocytes and chicken erythrocytes were intensely (Schiff's reagent for 60 min) or lightly Feulgen-stained (Schiff's reagent for 4 min). Integrated absorbances for the same 40 G0/G1 thymocytes and 20 chicken erythrocytes on slides from both staining groups were measured at different wavelengths of light. Figure 2 shows the integrated absorbances of the thymocytes and chicken erythrocytes plotted against the wavelengths of light at which the measurements were performed as well as the CVs for the integrated absorbances of the G0/G1 thymocytes at each wavelength.

The absorption maximum fo the Feulgen stain lies between 560 and 580 nm (Fig. 2). The CVs of the integrated

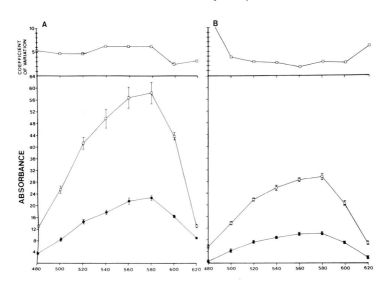

Figure 2. Integrated absorbances at different light wavelengths and intensities of Feulgen staining. Slides of mouse thymocyte/chicken erythrocyte mixtures were hydrolyzed and then incubated in Schiff's reagent for 4 or 60 min. Integrated absorbances were then measured for the same G0/G1 thymocytes and chicken erythrocytes on slides in each staining group at different wavelengths of light. (A) Standard Feulgen staining (60 min with Schiff's reagent). (B) Light Feulgen staining (4 min with Schiff's reagent). Off-peak measurements of the standard-stained cells in longer wavelengths ranges (600-620 nm) yielded lower CVs for the thymocyte integrated absorbances (A). Off-peak measurements of the lightly stained cells (at 480 nm and 620 nm) led to a marked increase in the CVs of the thymocyte integrated absorbances (B). (Reprinted by permission of publisher from "Increased Accuracy of Absorption Cytophotometric DNA Values by Control of Stain Intensity", by Allison et al, J Histochem Cytochem 29:1219-1228, 1981, copyright by the Histochemical Society, Inc.)

absorbances of the G0/G1 thymocytes on the intensely stained slide fell to approximately 3% when measured at 600 and 620 nm (Fig. 2A). The integrated absorbances of chicken erythrocytes on this slide were between 6 and 14 units in this wavelength range (Fig. 2A). The CVs were similar to those obtained for the integrated absorbances of lightly stained G0/G1 thymocytes (on slides with average chicken erythrocyte integrated absorbances also between 6 and 14 units) measured at 560 nm (Fig. 1B). The Figure also shows the deterioration in the CVs of the lightly stained G0/G1 thymocytes measured at 480 or 620 nm, wavelengths that produced very low average integrated absorbances for chicken erythrocytes (<6 units, Fig. 2B). This is analogous to the result obtained for very faintly stained chicken erythrocytes measured at 560 nm (Fig. 1B, average integrated absorbances of chicken erythrocytes <6 units). It is apparent that either altering of the staining reaction or selection of an off-peak wavelength of light for absorption measurements can be used for reduction of errors resulting from glare and diffraction (Allison et al, 1981).

Differences in local stain intensity. Nuclei often vary in average Feulgen stain intensity in different areas on the same slide (Atkin 1969). Thus, in order to obtain optimal estimates of cellular DNA content, one should ideally perform measurements of integrated absorbances on Feulgen-stained cells in one slide area. This requirement would seem to negate one of the major advantages of the absoprtion-cytophotometric technique: the ability to perform measurements on morphologically selected rare cells, which often requires scanning of large areas of a slide in order to find sufficient numbers of cells for study.

We found that avian and mammalian cells showed parallel changes in integrated absorbance when measured in different slide areas. Thus, the integrated absorbances of chicken erythrocytes can be used for correction of the effects of variations in local stain intensity over the slide area. Briefly, this "chicken erythrocyte transform" is performed as follows: A microcomputer averages the integrated absorbances of the chicken erythrocytes measured in a given slide area. All of the integrated absorbances measured for adjacent mammalian cells are then divided by the average integrated absorbance of the chicken erythrocytes. The resulting quotients are then multiplied by the cellular DNA content of a chicken erythrocyte (2.6 picograms/cell); this generates a cellular DNA value from each integrated absorbance value. This process is repeated for each separate slide area over which integrated absorbances are measured, and the cellular DNA values are collated in a single computer file (Allison et al, 1981).

The use of this correction is illustrated in Figure 3.

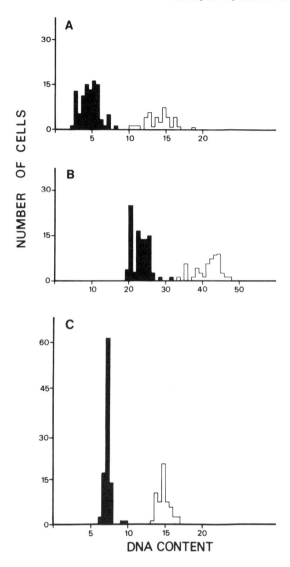

Figure 3. Measurement of DNA values for G0/G1 and mitotic thymocytes (M cells) at 560 nm on lightly and heavily stained slides. Integrated absorbances were measured in 4 central fields for G0/G1 cells and at all slide areas for M cells. Cellular DNA values were calculated with the erythrocyte transform. (A) Integrated absorbances of standard stained cells. The CV of the integrated absorbances of the G0/G1 cells were 9.1%, that for M cells was 11.0% and the

M:G0/G1 integrated absorbance ratio was 2.01:1. The erythrocyte transform applied to measurements made in different microscopic fields did not improve these CVs. (B) Integrated absorbances of lightly stained cells. The CV of G0/G1 integrated absorbances was 9.0%, that for M cells was 9.2%, and the M:G0/G1 integrated absorbance ratio was 2.02:1. (C) Erythrocyte transform of the integrated absorbances in (B). The CV of the DNA content of the G0/G1 cells was 2.8%, that for M cells was 4.0%, and the M:G0/G1 DNA ratio was 1.99:1. (Reprinted by permission of publisher from "Increased Accuracy of Absorption Cytophotometric DNA Values by Control of Stain Intensity", by Allison et al, J Histochem Cytochem 29:1219-1228, 1981, copyright by the Histochemical Society, Inc.)

Slides of a mouse thymocyte/chicken erythrocyte mixture were stained by the standard reaction (60 min Schiff's incubation) or were lightly stained (4 min Schiff's incubation). We measured the integrated absorbances for all thymocytes in four random fields, and for adjacent chicken erythrocytes in each of these fields, to obtain an estimate of the integrated-absorbance value at the G0/G1 peak. We then scanned all areas of the slides to find metaphase and anaphase cells (M cells). The integrated absorbances of each M cell and of 6 adjacent chicken erythrocytes were then measured. We obtained CVs of the integrated absorbances of 9.1% for G0/G1 cells and 11% for M cells, with an M/G1 integrated-absorbance ratio of 2.01:1, for cells stained by the standard reaction and measured at 560 nm (Fig. 3A). The erythrocyte transform did not improve the CVs of these absorption values. In Figure 3B, the integrated absorbances of nuclei of a lightly stained slide, measured at 560 nm, are shown; the CVs of the integrated absorbances of the G0/G1 and M cells were 9.0 and 9.2, respectively. The erythrocyte transform of these integrated absorbances (Fig. 3C) yielded a distinct improvement; the CVs for the G0/G1 and M cells decreased to 2.8 and 4.0, respectively, with an M to G0/G1 DNA ratio of 1.99:1. Similar results were obtained with the erythrocyte-transform DNA values for the off-peak measurements of standard-stained thymocytes: the CVs for the G0/G1 and M cells were 2.4 and 2.9, respectively, with a ratio of the M to G0/G1 DNA content of 2.00:1 (Allison et al, 1981).

<u>Errors due to insufficient stain darkness</u>. Nuclei with

insufficient stain darkness also yield artifactually low measured values of integrated absorbance. Although the exact reasons are unknown, ranges of optical density below which integrated absorbances cannot be measured accurately have been reported for several microdensitometric systems (Allison et al, 1981; Sklarew 1983; Swift, Rasch 1956). This effect is apparent in Figure 1A, where measurements of the integrated absorbances of faintly stained chicken erythrocytes at 560 nm gave relatively low values, as shown by a decreasing erythrocyte:thymocyte integrated absorbance ratio (on slides with chicken erythrocyte integrated absorbances averaging < 4 units). In Figure 2B, it can be seen that off-peak measurements, at 520 and 620 nm, of lightly stained G0/G1 thymocytes led to deterioration of the CVs of these integrated absorbance measurements on thymocytes. Thus it is necessary to test nuclei to determine whether they have sufficient stain darkness for accurate measurement of their integrated absorbances. A method for doing this is given below.

A method for determining the relationship between average nuclear point absorbance and errors in measurements of nuclear integrated absorbances for any image microdensitometer. Different types of G0/G1 mammalian white blood cells (WBC) have essentially identical DNA and Feulgen stain content after appropriate staining procedures have been applied (Allison et al, 1981; Allison et al, 1984; Bedi Goldstein 1976). Therefore, the integrated absorbances of WBC, which often vary in average stain intensity, can be used as references for an assessment of the relationship between average nuclear stain darkness and errors in integrated absorbance measurements for any microdensitometric system. The average nuclear point absorbance, which can be obtained manually from several point absorbance measurements over a nucleus with the scanning spot of the Vickers M5 microdensitometer, can yield an estimate of average nuclear stain darkness. In Figure 4, the average point absorbances and integrated absorbances of 15 G0/G1 mouse bone marrow cells, selected for heterogeneity in nuclear stain darkness, were measured at light wavelengths between 560 and 630 nm with a 100X achromatic lens and a 4% instruments glare level. Integrated absorbances were expressed in percent of the maximum value obtained for each light wavelength.

As shown in Figure 4, nuclei with very low (< 0.15) or high (> 0.70) point absorbances had relatively low integrated

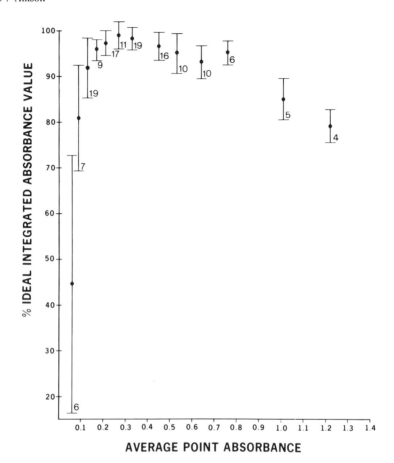

Figure 4. Relationship between the average point absorbance and errors in integrated-absorbance measurement. Integrated absorbances and point absorbances of Feulgen-stained G0/G1 mouse bone marrow cells were measured at wavelengths of light between 560 and 630 nm with a 4% instrument glare level. Integrated absorbances measured at each light wavelength are expressed as percentages of the maximum values obtained for nuclei measured at the same light wavelength. Integrated absorbances were decreased for nuclei with average point absorbances below 0.15 or above 0.70.

absorbance values. Therefore, to perform accurate measurements of integrated absorbances with this system, one would ideally select a wavelength of light, or a staining procedure if measurements were to be performed at the absorption maximum of the Feulgen stain, which places the average point absorbances of all nuclei to be measured between 0.15 and 0.70.

Thus, for any image cytometer that directly measures both integrated and point absorbances, one can determine the relationship between errors in integrated absorbance measurements and nuclear stain darkness by making a series of test measurements of point and integrated absorbances of Feulgen stained G0/G1 cells. Division of the integrated absorbance by the nuclear area will also give an estimate of the average nuclear stain darkness, thus allowing extension of this general method to microdensitometers that do not measure optical density directly (Allison et al, 1984). Furthermore, the experimentally determined relationship between nuclear stain darkness and errors in integrated absorbance for any microdensitometer can be stored in a microcomputer and used for automatic correction of integrated absorbances obtained for darkly stained nuclei.

Recommendations for integrated-absorbance measurements. Optical errors caused by instrument glare and diffraction are increased for darkly stained cells, i.e., relatively lower integrated absorbances are obtained for darkly than lightly stained nuclei. Therefore, decreasing the stain darkness, either with an actual reduction in the amount of stain in the nuclei (Fig. 1) or with the selection of an off-peak wavelength of light for the measurements (Fig. 2), can improve the accuracy of the measured integrated absorbance. Selecting an off-peak light wavelength is the simpler of the two procedures for reducing stain darkness and should be adequate for most experimental situations.

The extent to which stain darkness can be decreased is limited, however. At high transmission, the fraction of light absorbed by the chromophores is so small that it cannot be measured accurately, leading to low values of integrated absorbance for nuclei with insufficient stain darkness (Fig. 1, Fig. 2). Therefore, the wavelength of light selected must be sufficiently close to the absorption maximum of the Feulgen stain to allow accurate measurement of the most lightly stained nuclei present.

For any microdensitometric system, it is necessary to determine the range of optical densities over which accurate measurement of integrated absorbances can be performed

(Allison et al, 1981; Sklarew, 1983; Swift, Rasch 1956). Feulgen-stained G0/G1 WBC are suitable test objects for such measurements (Fig. 4). For the microdensitometric system employed in these experiments, the most lightly stained nuclei present must be measured at a light wavelength that give them an average point absorbance above 0.15. Some lightly stained nuclei must be measured at the absorption maximum of the Feulgen stain to meet this criterion. Indeed, some very lightly stained cells do not have sufficient stain darkness for accurate measurement even at the absorption maximum.

It may happen that, when an off-peak light wavelength is selected, or when an off-peak wavelength cannot be selected because of the presence of very faintly stained cells in a heterogeneously stained cell population, some of the nuclei are still too darkly stained for accurate determination of integrated absorbance (average point absorbance > 0.70 for the measuring system employed). In such cases, the integrated absorbance measurements can be corrected either by an electronic offset for instrument glare that can be applied to absorption cytophotometers which directly measure optical density (Goldstein 1970) or by a computer correction for stain darkness that can be applied to any image microdensitometer (Allison et al, 1984). Both of these procedures are straightforward and can improve the accuracy of integrated absorbances obtained for darkly stained nuclei.

However, it must be emphasized that, in most experiments, reduction of glare and proper selection of measuring light wavelengths will usually eliminate the need for these additional corrections for stain darkness. An example of the efficacy of this approach is given in Figure 5, which shows the DNA distribution, obtained from integrated-absorbance measurements, of 54 contiguous Feulgen-stained human WBC. These cells were stained under standard conditions, which produced average nuclear stain intensities ranging from light staining for monocytes to intense staining for small lymphocytes. The integrated absorbance measurements were performed at 620 nm with a 4% instrument glare level. This light wavelength was selected according to the criteria outlined above. No additional corrections for glare or stain darkness were performed. The CV of the measured integrated absorbances was 2.2% (Fig. 5), well within the range usually achieved by flow-cytometric measurements of the DNA content of human WBC.

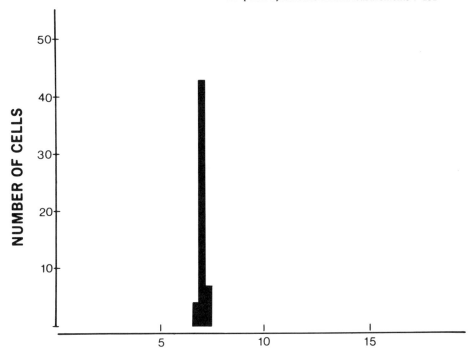

Figure 5. Measurement of the DNA content of Feulgen-stained human WBC at a single slide site. The integrated absorbances of 54 nuclei were measured at 620 nm with a 4% instrument glare level. The light wavelength was selected according to criteria outlined in the text. No further corrections of integrated-absorbance values for glare or stain darkness were applied. The CV of the integrated absorbances of the G0/G1 WBC are 2%.

Variability of Feulgen stain intensity in different sites on a slide can be partially overcome by the use of chicken erythrocytes as controls for local stain intensity and DNA content (Fig. 3). This solution, however, is not ideal, and the results are not as accurate as those for measurements of integrated absorbances of contiguous cells. This is because the absolute, rather than the relative, integrated absorbances of chicken erythrocytes at different slide sites must be compared. This comparison of absolute integrated absorbance values is not as accurate as is the measurement of the relative integrated absorbances of adjacent cells (Hiskey 1955). Therefore, slightly higher CVs are obtained for integrated-absorbance measurements of G0/G1 nuclei measured on diverse slide areas, even after the chicken erythrocyte transform, than when the same types of G0/G1 nuclei are measured in the one slide area.

Nevertheless, the degree of accuracy of integrated absorbance measurements, even after the erythrocyte transform, should be suitable for many biological and cell-kinetic experiments. For example, we conducted an experiment in which mouse bone marrow cells were labeled in vivo for 1/2 hr by an intravenous injection of {^{3}H}-thymidine (10 μgm/gm body weight, 43 Ci/mm, Amersham Searle, Arlington Heights, IL). After sacrifice, a bone marrow cell suspension was prepared, and chicken erythrocytes were added as a reference standard for local stain intensity. Slides were then prepared and Feulgen-stained. Next, the locations of G0/G1, S, and G2/M mouse cells and adjacent chicken erythrocytes were recorded over diverse slide areas on six separate slides. We measured the integrated absorbances of these cells at 588 nm, and we applied corrections for variations in stain intensity in different slide areas with the chicken erythrocyte transform. Autoradiographs of the slides were then prepared, and duplicate slides were exposed for 16, 32, and 64 days. Cells with 10 or more nuclear grains were considered to be labeled. The integrated absorbances obtained from the unlabeled G0/G1 cells on each slide, after correction with the erythrocyte transform, were set to average 7 pg DNA/cell by the microcomputer, and all other absorbances obtained on the same slide were adjusted accordingly. The DNA distributions of the labeled and unlabeled cells for all slides were collated into two computer files. The microcomputer automatically produced hard-copy graphic displays of these DNA distributions (Fig. 6).

Unlabeled cells were restricted to either G0/G1 or G2/M DNA content (Fig. 6A). The CV of the integrated absorbances of the unlabeled G0/G1 cells was 3.7%. All nuclei with apparent S-phase DNA content, as determined by the integrated absorbance measurements, were labeled (Fig. 6B). Thus, the cytophotometric measurements correctly predicted the incorporation of {^{3}H}-thymidine into nuclear DNA.

It is apparent that, if proper attention is paid to well-established methods of optimizing integrated-absorbance

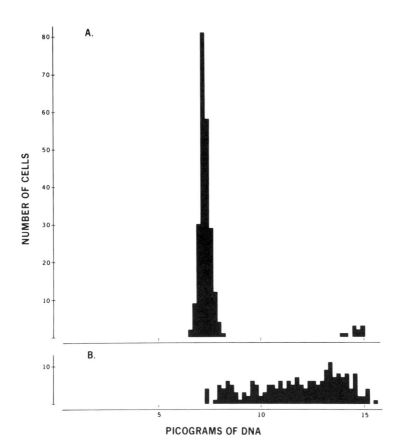

Figure 6. Measurement of the DNA content and {^{3}H}-thymidine incorporation of mouse bone marrow cells. The positions of 431 mouse bone marrow cells were recorded in different areas on six slides. Integrated absorbances were measured at 588 nm. The integrated absorbances of chicken erythrocytes were used as a reference for local stain intensity in different slides areas. Corrections for dark staining were performed automatically by a microcomputer (Allison et al., 1984). After absorption measurements, the slides were processed for autoradiography and exposed for 16 days or longer. The cells were then relocated and assessed for nuclear labeling. (A) DNA distribution of the unlabeled cells. (B) DNA distribution of the labeled cells. All S-phase were labeled (B). The CV of the integrated absorbances of the non-labeled G0/G1 cells was 3.7%.

measurements (Allison, et al, 1981; Allison et al, 1984; Duijndam et al, 1980; Goldstein 1970; Goldstein 1971; Goldstein 1975; Mayall, Mendelsohn 1970), absorption cytophotometry of Feulgen-stained cells can determine cellular DNA content with a degree of accuracy sufficient for most biological experiments. Although the instruments presently available do not allow as rapid a measurement of cellular DNA content for a large number of cells as is currently achieved with flow cytometry (Crissman, Tobey 1974), the absorption method would seem to be well suited for providing answers to many important biological questions. The major advantages of this technique are that it provides the capability of studying morphologically selected rare cells and of accurately measuring multiple parameters, in addition to DNA content, on the same cells. We are currently interfacing a computer-controlled stage to the Vickers M85 microdensitometer and developing the hardware and software necessary for sequential sets of measurements on the same cells. We hope that further developments in instrumentation, as well as future combinations of absorption cytophotometry with other techniques of image analysis, will lead to an increased understanding of the kinetics and mechanisms of various types of cell proliferation.

ACKNOWLEDGMENTS

I thank Mrs. Elisabeth F. Lanzl for her editorial assistance and Josala A. Fetherolf and Maxine Trujillo for their help in preparation of the manuscript.

LITERATURE CITED

Allison DC, Ridolpho P (1980). Use of a trypan blue assay to measure the deoxyribonucleic acid content and radioactive labeling of viable cells. J Histochem Cytochem 28:700-703.
Allison DC, Ridolpho P, Rasch E, Rasch R. Johnson T (1981). Increased accuracy of absorption cytophotometric DNA values by control of stain intensity. J Histochem Cytochem 29: 1219-1228.
Allison DC, Yuhas JM, Ridolpho PF, Anderson SL, Johnson TS (1983). Cytophotometric measurement of the cellular DNA content of {^{3}H}-thymidine-labeled spheroids. Cell Tissue Kinet 16:237-246.
Allison DC, Lawrence GN, Ridolpho PF, O'Grady BJ, Rasch RW, Rasch EM (1984). Increased accuracy and speed of absorption cytometric DNA measurements by automatic correction for

nuclear darkness. Cytometry 5:217-227.
Atkin NB (1969) Perimodal variation of DNA values of normal and malignant cells, Acta Cytol 13:270-273.
Bedi KS, Goldstein DJ (1976). Apparent anomalies in nuclear Feulgen-DNA contents. J Cell Biol 71:68-88.
Cooper EH, Peckham MJ, Millard RE, Hamlin ME, Gerald-Merchant R (1968). Cell proliferation in human malignant lymphomas. Eur J Cancer 4:287-296.
Crissman HA, Tobey RA (1974). Cell-cycle analysis in twenty minutes. Science 184:1297-1298.
Duijndam WAL, Smeulders AWM, Van Duijn P, Verweig AC (1980). Optical errors in scanning stage absorbance cytometry. I. Procedures for correcting apparent integrated absorbance values for distributional, glare, and diffraction errors. J Histochem Cytochem 28:388-394.
Fand SB, Spencer RP (1964). Off-peak absorption measurements in Feulgen cytophotometry. J Cell Biol 22:515-520.
Goldstein DJ (1970). Aspects of scanning microdensitometry. I. Stray light (glare). J. Microsc 92:1-16
Goldstein DJ (1971). Aspects of scanning microdensitometry. II. Spot size, focus, and illumination. J Microsc 93:15-42.
Goldstein DJ (1975). Aspects of scanning microdensitometry. III. The monochromator system. J Microsc 105:33-56.
Goldstein DJ (1982). Scanning microdensitometry of objects small relative to the wavelengths of light. J Histochem Cytochem 30:1040-1050.
Hisky CF (1955). Absorption spectroscopy. Phys Tech Biol Res 1:73-130.
Mayall BH, Mendelsohn ML (1970). Deoxyribonucleic acid cytophotometry of stained human leukocytes. II. The mechanical scanner of CYDAC, the theory of scanning photometry, and the magnitude of residual errors. J Histochem Cytochem 18:383-407.
Rasch EM, Rasch RW (1979). Applications of microcomputer technology of cytophotometry. J Histochem Cytochem 27:1384-1387.
Sklarew RJ (1983). Simultaneous Feulgen densitometry and autoradiographic grain counting with the Quantimet 720 D image-analysis system. J Histochem Cytochem 31:1224-1232.
Swift H (1950). The constancy of desoxyribose nucleic acid in plant nuclei. PNAS 36:643-653.
Swift H, Rasch E (1956). Microphotometry with visable light. Phys Tech Biol Res 3:353-400.
Tas J, van der Ploeg M, Mitchell JP, Cohn NS (1980). Protein staining methods in quantitative cytochemistry. J Microsc 119:295-311.

MICROPHOTOMETRIC DEMONSTRATION OF DIFFERENCES IN NUCLEAR STRUCTURE OR COMPOSITION.

Ronald R. Cowden and Sherill K. Curtis

Department of Biophysics
Quillen-Dishner College of Medicine
East Tennessee State University
Johnson City, Tennessee 37614

INTRODUCTION

For some time it has been recognized that quantitative cytochemistry, as applied specifically to nuclei and chromatin, can be used to investigate structural or compositional alterations in the nucleoproteins of chromatin which accompany changes in nuclear function. Bloch and Godman (1955) made the first serious attempts to use quantitative cytology to study changes in nuclear function, thereby demonstrating that the nuclei of regenerating diploid rat hepatocytes bound more methyl green (to DNA) and less pH 8.0-8.2 fast green (to histones) than metabolizing but nondividing diploid hepatocyte nuclei. The whole problem was first systematically examined by Ringertz (1969) who used a large number of methods to examine quantitative changes in nuclei, either in cells which were passing from a quiescent interphase condition to some form of activity, or in populations of cells changing from a condition of active proliferation to either a quiescent or a more differentiated state. In all instances diploid nuclei from the same species and sex were compared so that the amount of DNA per nucleus could be considered invariant, while the various reactions, including experiments with potentially quantitative staining methods, constituted the experimental variable.

It should also be noted in this context that the

introduction of "image-analysis" capabilities, generally but not exclusively based on detection of stained or fluorochromed nuclei with television cameras, and the subsequent reduction of digitized images have offered yet another approach to the objective analysis of the degree of condensation (i.e., "texture") of chromatin. Quantitative comparisons of the nuclei of Go and G_1 fibroblasts by Nicolini et al. (1977) and of the nuclei of diploid and tetraploid hepatocytes by Romen et al. (1980) are examples of the use of television technology. Mello (1983) used an X-Y stepping stage and a microphotometer with a photomultiplier tube to characterize textural changes in chromatin which had been stained by various chromatin-specific reactions.

With the advances in our knowledge of the molecular biology of chromatin, it would seem that the principal event in nuclei that relates to textural changes in chromatin, or to differences in nuclear protein, is transcriptional activity of DNA. Bonner (1979) estimated on the basis of the lability of chromatin to DNase I or DNase II that only approximately five percent of the total DNA of the genome is transcriptionally active in nuclei displaying highly condensed chromatin, while as much as twenty-five percent of the DNA is active in nuclei whose chromatin is more loosely organized. However, the percentage changes in dyebinding or fluorescence as chromatin becomes progressively less condensed have generally exceeded these percentages. Thus, it seems apparent that changes in nucleoprotein organization and composition are more complex than a simple reflection of an enhancement or restriction in transcriptional activity. While it may be possible at some time in the future to construct some correlations with transcriptional activity, these are not obvious--other than in a very general way--at this point in our exploration of chromatin and nucleoprotein structure.

Models of chromatin alteration that have been considered thus far (Ringertz, 1969) include mitogen-stimulated mammalian lymphocytes, hen erythrocyte nuclei incorporated into HeLa cell heterokaryons, mammalian spermatocytes, developing erythrocytes, regenerating and nonregenerating rat hepatocyte nuclei, cultured mammalian cells in logarithmic growth and in static phase, nuclei of larval and adult freshwater sponge archeocytes

(Harrison, Cowden, 1975) and the cells of the malphigian tubules of a South American blood-sucking insect subjected to feeding and starvation (Mello, 1983).

Cowden and Curtis (1981a,b; 1982a,b) have used isolated nuclei of mouse hepatocytes and small thymus lymphocytes (thymocytes) as models in studies of chromatin organization, since the former display largely "extended" chromatin while the latter contain a preponderance of highly condensed chromatin. While this approach differs from that of studies in which the same type of cell is examined in different developmental stages or functional conditions, this comparison appears to offer a valid model of extremes in terms of chromatin organization in nuclei presumably containing identical amounts of DNA. As indicated by Curtis and Cowden (1980), the basic approach has involved isolation of the nuclei by the hexylene glycol method of Wray et al. (1977), centrifugation onto cleaned slides held in special buckets resembling those designed by Lief et al. (1971), fixation of various types, fluorochroming by specific reactions under specified conditions, and measurement by microfluorometry using a Leitz MPV2 system (E. Leitz, Wetzlar, F.R.G.) configured as an incident light fluorometer. The microscope was interfaced to a PDP-8e minicomputer (Digital Equipment Corp.; Maynard, MA) supplied with CELANA software developed by Leitz. Nuclei were selected for measurement and focused upon in transmitted light using a Leitz NPL 25X planachromatic, oil immersion, phase-contrast objective. Measurements were obtained with the assistance of the computer which controlled the timing of the shuttering sequence and photometer readings. The microscope was equipped with a 150-watt xenon arc lamp, an epicondenser, and various combinations of excitation, dichroic, and barrier filters.

While fluorescence generally offers two to three orders of sensitivity above absorbance measurements (Ruch, 1966), it should be noted that most of the fluorescence procedures have absorbance counterparts and vice versa. In most cases, the absorbance methods are sufficiently sensitive to detect subtle differences in chromatin organization or significant alterations in proportions of nuclear proteins. While image-analysis methods are being developed for the evaluation of high-resolution fluorescence images, most of those currently in use were designed mainly for use in conjunction with absorbance images. Other physical-chemical methods have been used to study the higher-order structure of chromatin (Schmitz and Ramanathan, 1980; Nicolini et al., 1983; Zietz et al., 1983), but techniques of absorbance or fluorescence microphotometry represent direct extensions of cytology and can be conveniently directed toward a greater variety of material.

METHODS DIRECTED TOWARD DEMONSTRATION OF DNA

There are five absorbance dyes or categories of reactions selective for DNA that might be useful in quantitative cytochemistry or image-analysis: The Feulgen reaction, gallocyanin-chromalum, methyl green, pH 4.0 thiazin dyes (toluidine blue, Azures A and B, methylene blue, etc.) and the Immers et al. (1967) modification of the colloidal iron method. All of these except the Feulgen reaction require prior removal of RNA. Beyond these, ultraviolet absorbance at 260 nm is also useful, but it requires quartz optics. Rasch (1984) reports coefficients of percentage variation (C. V.'s) of about two percent in properly prepared nuclei stained with the conventional Feulgen reaction. Since nuclei displaying differing degrees of chromatin condensation can exhibit differences in absorbance of as much as thirty percent (Ringertz, 1969), the detection of small differences in chromatin structure is well within the capacity of state-of-the-art photometric systems. The gallocyanin-chromalum method (after RNase treatment) also tends to yield results that vary in nuclei whose chromatin displays differing degrees of condensation, while the methyl green method appears to be less useful for revealing differences in chromatin organization. As indicated by Mello (1983), pH 4.0 toluidine blue (after RNase treatment) displays either orthochromasia or metachromasia when it is bound, respectively, by extended or condensed chromatin. This dye can also produce selective birefringence when it is associated with condensed chromatin.

The modified colloidal iron method has been included in this group of reactions because it appears to be selective for DNA. Auer (1972) reported that the extended chromatin of blastogenic (mitogen-stimulated) lymphocytes binds more colloidal iron than the chromatin of unstimulated small lymphocytes. This latter is an example of enhanced binding of di- or tri-valent metal ions to entended chromatin, and the enhanced values obtained in similar preparations stained with the gallocyanin-chromalum method probably reflect the same mechanism.

New fluorescence methods should offer enhancements in sensitivity of two to three logarithmic orders, in part because the efficiency with which absorbed light is converted to fluorescence (i.e., the quantum efficiency) can be quite high when the dyes are bound by specific substrates. Among the nucleic acid-specific basic fluorochromes, several have been found to display a remarkable lack of sensitivity to differences in chromatin organization, as well as good linear stoichiometry; while others appear to be quite sensitive to differences in chromatin organization.

Probably more has been written about acridine orange than any other fluorescent dye (West, 1969; West and Lorincz, 1973), but the results obtained with this dye tend to be somewhat difficult to interpret because of their extreme dependence on the ratio of dye to potential binding sites in _all_ potential substrates (Nicolini et al., 1979; Darzynkiewicz, Traganos (1982); Kapuscinski et al., 1981; Darzynkiewicz et al., 1983). Furthermore, as Darzynkiewicz et al. (1983) pointed out, at certain higher concentrations, acridine orange can denature and form precipitation complexes with nucleic acids. This occurs with single-stranded nucleic acids at lower concentrations than with double-stranded nucleic acids. Nevertheless, if the concentration of the dye is adjusted properly, microspectrofluorometric measurements of green and red emissions can be used as estimates, respectively, of double-stranded and single-stranded nucleic acids. If condensed chromatin produces yellow or golden emission colors, the concentration of dye is too high to allow absolute measurement of nucleic acids. However, a higher dye concentration can be selected deliberately to demonstrate and emphasize differences in chromatin organization. Flow cytophotometric studies conducted by Belmont and Nicolini (1983) are excellent examples of the manipulation of acridine orange by a methods developed by Darzynkiewicz and Traganos (1982) to obtain data concerning chromatin structure. In common with some other basic metachromatic dyes, acridine orange forms intercalating complexes with double-stranded nucleic acids and electrostatic complexes with single-stranded nucleic acids (as well as with acidic mucins, some acidic phospholipids or polyphosphates as in "volutin" granules of yeast, or with carboxylic acids of proteins). As Weisblum (1973) pointed out, the quantum efficiency of acridine orange is enhanced by its association with either A-T or G-C base pairs.

There are a number of Hoechst compounds which include Hoechst 33258, 33343 and "Nuclear Yellow" which can be used at neutral pH to demonstrate DNA selectively. These, along with DIPI and DAPI, are A-T selective (i.e., their quantum efficiencies are enhanced only by associations of the dyes with A-T base-pairs.) The Hoechst reagents are bound by a nonintercalating mechanism (Krey, 1980). While at neutral pH they display only trivial affinity for RNA, Hoechst 33258 and DIPI fluorochrome both DNA and RNA at acidic pH levels (Hilwig and Gropp, 1975; Curtis and Cowden, 1981; 1983). Furthermore, at neutral pH they are relatively insensitive to chromatin structure, while at an acidic pH after RNase pretreatment they become probes of chromatin organization.

Mithramycin, Chromomycin A_3, and Olivomycin are nonintercalating fluorochromes whose binding and fluorescence require the presence of G-C base-pairs. When used under the conditions recommended by Johannisson and Thorell (1977), they are relatively insensitive to differences in chromatin organization and do not demonstrate RNA (Cowden and Curtis, 1981a). However, if the preparations are pretreated with RNase (Swift, 1966), there is a large but unexplained increase in fluorescence which is greater in thymocyte nuclei than in hepatocyte nuclei (Cowden and Curtis, 1981a). Since samples of Mithramycin obtained recently from different sources can produce different results and probably do not have the same composition, most experiments performed since 1982 have been undertaken using Chromomycin A_3 whose binding properties and fluorescence are very similar to those of Mithramycin.

Enhancement of fluorescence after treatment with RNase can also occur in material stained with a fluorescent analog of actinomycin D, 7-aminoactinomycin D. As indicated by Cowden and Curtis (1981a), this fluorochrome is relatively insensitive to differences in chromatin organization when it is used a neutral pH, but the total fluorescence of nuclei can increase by as much as 200 percent after treatment with RNase. Although such dramatic increases in the fluorescence of material stained with 7-aminoactinomycin D and other G-C selective fluorochromes cannot be explained fully at present, it seems unlikely that increases of this magnitude can be produced by simple enzymatic removal of RNA. Since recent experiments by Bendayan (1981) involving the localization of RNase labelled with colloidal gold revealed that RNase is bound preferentially by condensed chromatin, it is possible that enhancement in fluorescence obtained with G-C-selective fluorochromes might depend in some way on RNase-induced alterations at sites coinciding with transcriptionally inactive portions of the genome. However, additional work is required to confirm such a mechanism or to uncover others.

Proflavine (3,6-diaminoacridine) is unusual since it is believed to react with A-T and G-C base pairs through, respectively, nonintercalating and intercalating mechanisms. As indicated by Thomes et al. (1969), the association of proflavine with A-T base-pairs produces fluorescent products, while its interaction with G-C base-pairs tends to reduce fluorescence. When proflavine was used to stain isolated hepatocyte and thymocyte nuclei, Cowden and Curtis (1981a) found that pretreatment with RNase produced an enhancement in fluorescence resembling that obtained with G-C dependent dyes. As in the case of the latter, the mechanisms underlying such enhancement are not known at present, but it is possible that preferential binding of RNase to condensed chromatin might be involved in some way.

Of the DNA-selective fluorochromes studied, quinacrine mustard, which is expected to interact with DNA at sites of A-T base-pairs, is particularly sensitive to differences in chromatin organization. As indicated by Moser and Meiss (1982), Moser and Müller (1979), and Moser et al., (1975), quinacrine can be used to demonstrate changes in chromatin organization that accompany the cell cycle, including the condensation occurring in G_1 immediately before S-phase. Such chromatin condensation during pre-S-phase was also revealed by Belmont and Nicolini (1983) with the use of low concentrations of acridine orange. In other studies,

Ferrucci and Mezzanotte (1982) demonstrated that photo-oxidative destruction of guanine residues, induced by exposure to a tungsten lamp in the presence of 10^{-4} M methylene blue, both increased quinacrine fluorescence and extended fluorescence into chromosome regions that were previously nonfluorescent. Their findings are consistent with the reports of Jovin et al. (1979) that quinacrine fluorescence is dependent on the presence of A-T bases and that the fluorescence can be influenced by the presence of guanine residues. In a study involving X-ray microanalysis of chloride in mammalian metaphase chromosomes, Sumner (1981) demonstrated that quinacrine is bound uniformly along the length of these chromosomes. Thus, the characteristic fluorescent banding patterns obtained with this compound (Caspersson et al., 1969) appear to be related to differences in quantum efficiency of the dye at different sites rather than to uneven binding of the dye along the chromosomes. Comings and Drets (1976) proposed that associations of nonhistone chromosomal proteins are responsible for these differences. This may also be true in the case of the interaction of quinacrine compounds with DNA in interphase nuclei, since, as indicated by Cowden and Curtis (1981a), quinacrine mustard was the only DNA-selective fluorochrome in a group of nine that produced higher total fluorescence values in thymocyte nuclei than in diploid hepatocyte nuclei, and which displayed a failure in stoichiometry with 4c and 8c hepatocyte nuclei. In addition to any inherent affinities for specific base-pairs the quinacrines may exhibit, they are extremely sensitive to differences in chromatin organization which, in turn, are probably dependent on proteins associated with chromatin (Bhorjee et al., 1983).

Malinin (1978) found that aluminum ions bind selectively to nucleic acids in spreads, squashes, and smears (but not in sections). When exposure to aluminum is followed by reaction with the fluorescent chelating agent "morin," it is possible to visualize the distribution of aluminum ions. This fluorescence is extremely stable, and permanent preparations can be made which survive over a period of years without any apparent changes. However, as in the case of iron, aluminum and indium ions (both selective for nucleic acids under the conditions specified) function as probes of higher-order chromatin organization, and the more open or extended chromatin configuration of mouse hepatocyte nuclei binds more of these ions than chromatin in condensed thymocyte nuclei (Cowden and Curtis, 1981b). Other ions which form complexes with morin, including zinc, produce lower values by at least a logarithmic order and do not appear to be useful in studies of chromatin organization.

There are other DNA- or nucleic acid-specific fluorochromes that may be useful as probes of chromatin organization. Propidium iodide and ethidium bromide intercalate into double-stranded nucleic acids and, in common with acridine orange, become associated nonselectively with both G-C and A-T base-pairs (LePecq and Paoletti, 1967; Hudson et al., 1969; Waring, 1970). Mazzini et al. (1983) demonstrated that propidium iodide can be used at very low dye-DNA ratios as a successful probe of differences in chromatin stucture. Berberine sulfate can also be used to demonstrate nucleic acids, and, as indicated by Moutschen (1976), it is particularly effective as a chromosome-banding agent. Pyronin Y is a G-C selective, intercalating dye, but it displays a distinct preference for single-stranded nucleic acids over double-stranded nucleic acids (Cowden and Curtis, 1981a). Berberine sulfate and pyronin Y do not appear to offer any specific advantages in studies of chromatin over some of the other dyes that have been discussed. On the other hand, adriamycin and daunomycin (and related compounds) might prove to be especially useful since they display a preference for segments or blocks of DNA in which A-T base-pairs form uninterrupted sequences of various lengths. Thus, these compounds do not appear to be capable of providing estimates of total A-T content of chromatin, but rather they seem to demonstrate sites at which A-T base-pairs form highly repetitive segments.

There are a number of dyes that either have been used in survey experiments (Schwartz and Wittekind, 1982; Popov and Thorell, 1982) or have not been entered formally into the literature as nucleic acid-selective fluorochromes that might prove to be useful as probes of chromatin structure if they display a tendency to bind in a nonstoichiometric fashion. The usefulness of these compounds can be assessed only after additional careful study.

Fluorescent versions of the Feulgen reaction for DNA are possible in which a fluorescent Schiff reagent (Ruch, 1966) or pseudo-Schiff reagents (Böhm and Fukuda, 1981) are substituted for a conventional Schiff reagent. However, Bjelkenkrantz (1983) demonstrated by using very sensitive microfluorometry that the C.V.'s obtainable with one of the more reliable pseudo-Schiff reagents in the fluorescent Feulgen reaction are about six to seven percent at the lowest, while the C.V.'s of approximately two percent can be obtained routinely in preparations fluorochromed with Hoechst 33258. This limits the value of the fluorescent Feulgen reaction in studies requiring very accurate estimates of DNA, but the method is sufficiently accurate to demonstrate differences in ploidy. Furthermore, as indicated by Marks (1983) and Allison (1984), the Feulgen reaction can be manipulated to demonstrate specific subsets of nucleoprotein by over- or under-hydrolysis, or by manipulation of the staining conditions. Mello (1983), for example, employed over-hydrolysis in 0.1N HCl at 37°C. for the selective demonstration of more highly condensed chromatin. Equivalent results should be possible with the fluorescent Feulgen reaction, and the resulting preparations should be particularly suitable for use in image-analysis systems.

There are useful methods by which DNA may be manipulated before it is stained and examined. Specific DNA-protein complexes display different points at which they are denatured by heat. Denaturation of pure DNA occurs at a lower temperature than that of nucleoprotein complexes, some of which may not denature until temperatures of between 80 and 90°C. are reached. Ringertz (1969) used staining with acridine orange to demonstrate the effects of thermal denaturation by measuring green and red fluorescence, attributed to, respectively, undenatured (double-stranded) and denatured (single-stranded) nucleic acids. These values could be compared subsequently with absorbance of ultraviolet light at 260 nm which displays a hyperchromic shift after thermal denaturation (Kernell and Ringertz, 1972). Barni et al. (1981) substituted propidium iodide for acridine orange in order to obtain emissions of a single color. In a comparative study of the nuclei of neutrophils and lymphocytes, Dreskin and Mayall (1974) studied reductions in staining with gallocyanin-chromalum that occurred when the nuclei were subjected successively to thermal denaturaton and treatment with S_1 nuclease, which is selective for single-stranded nucleic acids. The effects of thermal denaturation have also been detected from losses in polarized fluorescence of preparations stained with acridine orange (McInnes and Uretz, 1966, 1967). In general, nuclei containing considerable amounts of condensed chromatin display more resistance to thermal denaturation than nuclei whose chromatin is loosely organized. Similarly, denaturation-renaturation might be used to establish the proportion of unique (slowly reassociating) versus intermediate and highly repetitive sequences in a genome. Stockert (1972) initially attempted this using acridine orange to discriminate between double-stranded and single-stranded nucleic acids, while Moritz (1977) made use of this approach to study the process of chromatin diminution in *Ascaris*. While the approach has not been widely used or critically compared to molecular biological approaches, it does offer another quantitative cytochemical method for partial characterization of the structure of the genome.

METHODS DIRECTED TOWARD DEMONSTRATION OF PROTEINS

The options for discriminations among nuclei using cytochemical reactions directed toward the demonstration of proteins are more limited than those available for the demonstration of DNA. Methods designed for the demonstration of histones would immediately come to mind because histones, as an integral part of the nucleosome complex in chromatin, are metabolically stable and present at a fixed ratio to DNA. Both absorbance (Cohn, 1973) and fluorescence (Cowden and Curtis, 1982a) versions of the Alfert and Geschwind (1953) method for demonstrating histones result in failures in stoichiometry that qualify these as probes of chromatin organization rather than as methods for demonstrating absolute amounts of histones per nucleus. This is also the case for the ammoniacal silver reaction of Black and Ansley (1964) and the modified Guard reaction as presented by Cowden et al. (1976). The fluorescent versions of the Guard procedure developed by Cowden and Curtis (1976) can be used to demonstrate either all basic groups in chromatin or only those found in more condensed chromatin. The Guard reaction represents a variation of the "big dye-little dye" principle of Horobin (1982) in which only a single acidic dye was used in the variation, and in which differentiation with polyacid was employed. The effect of differentiation was to displace dye first from general cytoplasmic structures, then from extended chromatin, and finally from condensed chromatin.

Thus, the results obtained depend on the investigator's judgement concerning the differentiation step. It is not as selective as either of the other histone methods, but in common with them, the staining is abolished by prior extraction of preparations with dilute mineral acids or by reactions that block amino groups. This reaction can produce results similar to those obtainable with the long-hydrolysis Feulgen reaction (Mello, 1983), but it is directed toward demonstration of proteins rather than DNA.

There is yet another method which was developed by Barnard and Danielli (1956); and Barnard (1961), and simplified by Curtis and Cowden (1970) which results in the selective demonstration of chromatin by a color reaction with protein end-groups. The method involves freeze-drying, freeze-substitution or chemical fixation using cold ethanol-acetone (1:1), subsequent anhydrous benzoylation, hydration, and demonstration of histidine residues with a diazonium hydroxide. The premise of this reaction is that the existence of a complex between nucleic acids and proteins, or between proteins and glycosaminoglycans protects some of the histidine residues from the effects of anhydrous benzoylation. Following hydration of these protected histidine residues, swelling occurs and they can be revealed by treatment with diazonium hydroxides. Curtis and Cowden (1970) used Fast Black K salt for this purpose. While the method was not expected to demonstrate any specific subset of proteins, extraction of nuclei with dilute mineral acids before fixation prevented the reaction. Cowden (1974) reported that integrated absorbance values for nuclei of bullfrog erythroblasts are analogous to the pattern one obtains with DNA measurements since G_1, S and G_2 phases can be recognized. Comparisons between the integrated absorbance of mature nucleated erythrocytes and polymorphonuclear leucocytes indicated that modal values were greater in the latter, so that the method can be considered as a reasonable indicator of chromatin organization. Unfortunately, it has not been possible to substitute any of the fluorescent diazonium salts listed in Pearse (1968) in this reaction, so at this time it can only be used as an absorbance reaction.

Smetana and Busch (1966) developed a sequence using pH 9.0 toluidine blue O after formalin fixation and extraction of preparations with 5 percent trichloroacetic acid (TCA) followed by a Feulgen-type hydrolysis in 1N HCl at $60°$ C.

These treatments were designed to remove successively nucleic acids and some formaldehyde crosslinks, then histones and possibly a few other proteins. The pH 9.0 basic dye was then reacted with ionized carboxyl groups of the remaining nucleoproteins. The staining pattern obtained in nuclei is strikingly similar to some contemporary demonstrations of "nuclear matrix" (Kaufmann et al., 1981). Cowden and Curtis (1984-in press) modified the reaction by using isolated nuclei, by extracting histones with $0.4N$ H_2SO_4 before fixation in formalin, and by staining at pH 9 with basic fluorescent dyes (acridine orange, berberine sulfate, rivanol). Microfluorometric evaluation of these preparations indicated that extraction with H_2SO_4 before fixation produced dramatic increases in fluorescence whose absolute magnitude varied from approximately 200 to 300 percent, depending on the basic fluorochrome used in the procedure. As expected, diploid hepatocyte nuclei produced higher total nuclear fluorescence values in all of the categories that were compared than thymocyte nuclei. The original method was also used by Desai and Foley (1974) to compare normal lymphoid cell nuclei with those of lymphoid tumor cells. The tumor cells produced higher modal values than the normal lymphoid cells. Of the nucleoprotein-specific methods employed, this one produced the most obvious morphological discrimination between normal and tumor nuclei.

The total protein content of nuclei also offers some reflection of metabolic activity, and measurements of changes in protein content by microinterferometry have been routinely included in investigations from the Karolinska Institutet group (Ringertz, 1969). However, absorbance microspectrophotometric investigations have been complicated (Frederiks, et al., 1980) by the assumption that histones probably do not stain at pH 2.8 with acidic dyes, as suggested by Deitch (1955). Subsequent studies by Deitch (1965), Leemann and Ruch (1972) and Cowden and Curtis (1982b) have demonstrated that this is not the case. Total histone per nucleus can be estimated by comparing unextracted nuclei with those from which the histones have been removed by dilute acid. As demonstrated by Cowden and Curtis (1982b) both diploid hepatocyte and thymocyte nuclei display virtually identical reductions in absolute units of total nuclear fluorescence after extraction with $0.4N$ H_2SO_4 which would be expected if both kinds of nuclei contained the same amount of DNA and the same ratio of DNA to histones. Treatment of preparations after fixation with hot five percent TCA both extracts nucleic acids and removes some formaldehyde crosslinks, thereby increasing the number of available binding sites and reducing variation. By use of hot TCA on otherwise unextracted nuclei, it was possible to separate diploid hepatocyte nuclei from diploid nonhepatocyte nuclei--chiefly Kupffer and endothelial cell nuclei--by microfluorometric measurements. Use of 0.35M NaCl, which removes most easily soluble and lightly bound proteins, including the LMG and HMG groups of proteins, also reduced variability within given nuclear types and increased the measured differences between diploid hepatocyte and thymocyte nuclei. These values were spread even further when both histones and nucleic acids were extracted.

CONCLUDING REMARKS

The methods used in these investigations and the reference sources for DNA are given in Table 1, and for proteins in Table 2. The formal data were presented in the original publications cited.

It would seem reasonable, given the high state-of-the-art in the molecular biological characterization of nucleic acids and the detection of specific proteins, to question why these relatively nonspecific methods would seem to be useful in contemporary cell biology. The argument has to be split into two parts. In automated diagnostic cytology, the objective is reduced to separation of cell or nuclear categories by some machine-sensible signal. The cells of various classifications must fall within certain nonoverlapping boundary values; or if they do overlap, other parameters must be measurable so that a very high percentage of cells can be assigned ultimately to some finite sets or categories, or declared pathological. Most of the methods discussed in this presentation have potential application in this area, either used alone or in combination with other methods. In basic cell biology, one needs only to reflect on the total resources required to characterize DNA by restriction endonuclease mapping, cloning in plasmids, etc., or the problems inherent in the preparation of monoclonal or polyclonal antibody to a small sample to realize that this is an expensive and refined

TABLE 1

METHODS FOR DEMONSTRATION OF DNA IN NUCLEOPROTEIN

DYE/METHOD	SPECIFIC FOR	PROBE OF CHROMATIN STRUCTURE	CITATION(s)
Standard Feulgen	Apurinic Acid	Yes, can be manipulated	Marks (1983) Mello (1983)
*Fluorescent Feulgen	Apurinic Acid	Yes, can be manipulated	Böhm, Fukudka (1981)
Methyl Green	A-T Bases, intercalates into ds-DNA	Yes, intercalates into ds-DNA	Diaz (1972)
Gallocyanin-Chromalum	Nucleic Acids, Gen.	Yes, binding of chromealum	Dreskin & Mayall (1974)
Toluidine Blue O pH 4.0	Nucleic Acids, Gen.	Yes, metachromasia, Birgfrengence	Mello (1983)
*Hoechst 33258, 33342	A-T Bases, non-intercalating	Only at acid pH	Gropp & Hilwig (1975)
DAPI, DIPI	A-T Bases, non-intercalating	Only at acid pH	Curtis, Cowden (1981, 1983)
Mithramycin, Chromomycin A_3, Olivomycin	G-C Bases, non-intercalating	Only after RNase pre-treatment	Cowden, Curtis (1981a)
*Proflavine 2HCL	A-T Bases, (non-intercalating) fluoresence enhanced; G-C Bases (intercalating and quinches)	Yes, after RNase pre-treatment thymocyte fluoresence increases more than in hepatocyte nuclei	Cowden, Curtis (1981a)
*Quinacrine 2HCL	A-T Bases, intercalates	Sensitive to chromatin condensation; selective for A-T Bases and sensitive to guanine insertion	Moser, Miess (1982) Cowden, Curtis (1981a) Ferucci, Mezzonotte (1982)
*Berberine Sulfate	Nucleic Acids, Gen.	Produces chromosome bands	Moutschen (1976) Cowden, Curtis (1981a)
Colloidal Iron	Nucleic Acids, Gen.	Yes	Auer (1972)
*Al^{+3}, In^{+3}+Morin	Nucleic Acids, Gen.	Yes, only in isolated cells	Cowden, Curtis (1981b)
*Propidium Iodide, Ethidium Bromide	ds-Nucleic Acids, prefers Pyrimidine-Purine Bases	Only at very low Dye/DNA ratios	Mazzini et al. (1983)
Acridine Orange	ds-DNA (Green) SS-Nucleic Acid (Red)	Depends on concentration ratios; excess of dye alters emission colors	Nicolini, et al (1979) Darzynkiewicz, Tragonos (1982)
Absorbance at 260 nm	Nucleic Acids, Gen.	Displays Hyperchromasia upon denaturation, depolymerization	Kernell, Ringertz (1972)

*Fluorescent

References not given in body of paper

Cowden RR, Curtis SK (1983). Supravital experiments with Pyronin Y, A fluorochrome of mitochondria and nucleic acids. Histochem. 77, 535.

Diaz, M (1972). Methyl Green staining of highly repetitive DNA in polytene chromosomes. Chromosoma (Berl.) 37, 131.

technology that has, with rare exceptions, been practiced on a restricted selection of "standard" and universally accepted models: human and mouse cells in culture, Xenopus laevis, Drosophila melanogaster and various species of sea urchins. It is improbable that this kind of effort or costly resources will be expended on the obscure or the unusual species, or the special kind of cells that have so frequently offered some of the more interesting insights into cell and developmental biology in the past. While the investment in the instrumentation involved in this kind of research is not trivial, once the equipment is available, it allows a far broader survey of material than would be open to the methods of contemporary molecular biology. To cell and developmental biologists, these methods offer a capacity to screen systems for major differences in nuclear organization or composition.

TABLE 2

METHODS FOR DEMONSTRATION OF NUCLEAR PROTEINS

DYE/METHOD	SPECIFIC FOR	PROBE OF CHROMATIN STRUCTURE	CITATION(s)
Alkaline Acid Dye	Histones	Yes	Bloch, Godman (1955) Cowden, Curtis (1982a)
Ammonical Silver	Histones	Yes	Black, Ansley (1964) Cowden, Curtis (1976)
Guard Method	Probably Histones	Yes, can be manipulated	Cowden, Curtis (1976)
Benzoylation-Diazonium	"Protected" Histidine Residues, Proteins Tightly bound to DNA	Yes	Barnard, (1962) Curtis, Cowden (1970)
Basic Dye, pH 9.0	Requires removal of Histones and Nucleic Acids	Yes	Smetana, Busch (1966) Cowden, Curtis (1984)
Basic Groups of Protein	This is a "Total-Protein" Method	Yes, improved by use of hot 5% TCA, and extractions	Cowden, Curtis (1982b)

NOTE: Absorbance and fluoresence versions are available for all these reactions except benzoylation-diazonium and ammonical silver.

LITERATURE CITED

Alfert M, Geschwind I (1953). A selective staining method for basic proteins of cell nuclei. Proc Nat Acad Sci (USA) 39: 991.

Allison D (1984). Refinements in absorption microspectrophotometric analysis of chromatin and nuclei. Advances in Microscopy.

Arndt-Jovin DJ, Latt SA, Striker G, Jovin TM (1979). Fluorescence decay analysis in solution and in a microscope of DNA and chromosomes stained with quinacrine. J Histochem Cytochem 27:87.

Auer G (1972). Cytochemical properties of nuclear chromatin as demonstrated by the colloidal iron binding technique. Exp Cell Res 75: 237.

Barnard EA (1961). Acylation and diazonium coupling in protein cytochemistry with special reference to the benzoylation-tetrazonium method. In: General Cytochemical Methods (JF Danielli, ed) Vol. 2, pp 203-258. Academic Press. 258. Academic Press, New York.

_____, Danielli JF (1956). A cytochemical reaction for nucleoprotein. Nature (Lond.) 67: 219.

Barni S, De Piceis Polver P, Gerzeli G, Nano R (1981). Propidium iodide as a probe for the study of chromatin thermal denaturation in situ. Histochem J 13: 781.

Belmont A, Nicolini C (1983). The G_1 period. Two cycles of chromatin conformational changes monitored by single cell dye intercalation. Cell Biophys 5: 79.

Bendayan M (1981). Electron microscopical localization of nucleic acids by means of nuclease-gold complexes. Histochem J 13: 699.

Bhorjee JS, Barclay S, Wedrychowski A, Smith AM (1983). Monoclonal antibodies specific for tight-binding human chromatin antigens reveal structural rearrangements within the nucleus during the cell cycle. J Cell Biol 97: 389.

Bjelkenkrantz K (1983). An evaluation of Feulgen-acriflavine-SO_2 and Hoechst 33258 for DNA cytofluorometry in tumor pathology. Histochem 79: 177.

Black MM, Ansley HR (1964). Histone staining with ammoniacal silver. Science 143: 693.

Bloch DP, Godman GC (1955). Evidence of differences in the desoxyribonucleoprotein complex of rapidly proliferating and non-dividing cells. J Biophys Biochem Cytol 1: 531.

Böhm N, Fukuda M (1981). Fluoreszierende Farbstoffe vom Schiff-typ. Acta Histochem Suppl 24: 181.

Bonner J (1979). Expressed and unexpressed portions of the genome: their separation and characterization. In: Chromatin Structure and Function (C Nicolini, ed) Part A, pp 15-23. Plenum Press, New York.

Caspersson T, Zech L, Modest EJ, Foley GE, Wagh U, Simonsson E (1969). DNA-binding fluorochromes for the study of the organization of the metaphase nucleus. Exp Cell Res 58: 141.

Cohn NS (1973). A model system analysis of the parameters in histone staining: I. Alkaline fast green. Histochem J 5: 529.

Comings DE, Drets ME (1976). Mechanisms of chromosome banding. IX. Are variations in DNA base composition adequate to account for quinacrine, Hoechst 33258 and daunomycin banding? Chromosoma (Berl.) 56: 199.

Cowden RR (1974). Mikrospectrophotometrische Untersuchungen an der Interphasischen Chromosalen Nucleoprotein-Struktur. Verh Deutsch Zool Ges 1974: 130.

_____, Curtis SK (1975). A comparison of four quantitative cytochemical methods directed toward demonstration of DNA. Histochem 45: 299.

_____, _____ (1976). Cytochemical characterization of the modified Guard procedure, a regressive staining method for demonstrating chromosomal basic proteins. II. Substitution of dyes for biebrich scarlet. Histochem 48: 93.

_____, _____ (1981a). Microfluorometric investigations of chromatin structure. I. Evaluation of nine DNA-specific fluorochromes as probes of chromatin organization. Histochem 72: 11.

_____, _____ (1981b). Microfluorometric investigations of chromatin structure. II. Mordant fluorochroming with ions that complex with morin. Histochem 72: 391.

_____, _____ (1982a). Microfluorometric investigations of chromatin structure. III. Estimation of histones and DNA in thymocyte and hepatocyte nuclei. Effects of extraction at pH 3.0 Histochem 74: 469.

―――――――, ――――――― (1982b). Microfluorometric investigations of chromatin structure. IV. Determination of total protein values in thymocyte and hepatocyte nuclei. Effects of extraction with 0.4N H_2SO_4 and 0.35M NaCl. Histochem 74: 329.

―――――――, Rasch EM, Curtis SK (1976). Cytochemical evaluation of the Guard procedure, a regressive staining method for demonstrating chromosomal basic proteins. I. Effects of fixation, blocking reactions, selective extractions, and polyacid "differentiation." Histochem 48: 81.

Curtis SK, Cowden RR (1970). A simplified benzoylation-diazonium coupling sequence for demonstrating conjugated proteins. Histochem 23: 7.

―――――――, ――――――― (1980). Effects of preparation and fixation on three quantitative cytochemical procedures. Histochem 68: 29.

―――――――, ――――――― (1981). Four fluorochromes for the demonstration and microfluorometric estimation of RNA. Histochem 72: 39.

―――――――, ――――――― (1983). Evaluation of five basic fluorochromes of potential use in microfluorometric studies of nucleic acids. Histochem 78: 503.

Darzynkiewicz Z, Traganos F (1982). RNA content and chromatin structure in cycling and noncycling cell populations studied by flow cytometry. In: Genetic Expression in the Cell Cycle (GM Padilla, KS McCarty, eds), pp 103-128. Academic Press, New York.

―――――――, Evenson D, Kapuscinski J, Melamed MR (1983). Denaturation of RNA and DNA in situ induced by acridine orange. Exp Cell Res 148: 31.

Deitch AD (1955). Microspectrophotometric study of the binding of the anionic dye, naphthol yellow S by tissue sections and by purified proteins. Lab Invest 4: 324.

――――――― (1965). A cytophotometric method for the estimation of histone and non-histone protein. J Histochem Cytochem 13: 17.

Desai LS, Foley GE (1974). Human leukemic cells. Cytochemical studies on acidic nuclear proteins. J Histochem Cytochem 22: 40.

Diaz M (1972). Methyl green staining and highly repetitive DNA in polytene chromosomes. Chromosoma (Berl.) 37:131.

Dreskin SC, Mayall BH (1974). Deoxyribonucleic acid cytophotometry of stained human leucocytes. III. Thermal denaturation of chromatin. J Histochem Cytochem 22: 120.

Ferrucci L, Mezzanotte R (1982). A cytological approach to the role of guanine in determining quinacrine fluorescence response in eukaryotic chromosomes. J Histochem Cytochem 30: 1289.

Frederiks WM, Slob A, Schroder M (1980). Histochemical determination of histone and non-histone protein content in rat liver nuclei. Histochem 68: 589.

Harrison FW, Cowden RR (1975). Feulgen microspectrophotometric analysis of dexoyribonucleoprotein organization in larval and adult freshwater sponge nuclei. J Exp Zool 193: 131.

Hilwig I, Gropp A (1975). pH-dependent fluorescence of DNA and RNA in cytologic staining with "33258 Hoechst." Exp Cell Res 91: 457.

Horobin RW (1982) Histochemistry: An Explanatory Outline of Histochemistry and Biophysical Staining. Gustav Fisher Verl., Stuttgart-New York.

Hudson B, Upholt WB, Devinny J, Vinograd J (1969). The use of an ethidium analogue in the dye-buoyant density procedure for the isolation of closed circular DNA. Proc Natl Acad Sci USA 62: 813.

Immers J, Markman B, Runnström J (1967). Nuclear changes in the course of development of the sea urchin studied by means of Hale staining. Exp Cell Res 45: 425.

Johannisson E, Thorell B (1977). Mithramycin fluorescence for quantitative determination of deoxyribonucleic acid in single cells. J Histochem Cytochem 25: 122.

Kapuscinski J, Darzynkiewicz Z, Melamed MR (1982). Luminescence of the solid complexes of acridine orange with RNA. Cytometry 2: 201.

Kaufmann SH, Coffey DS, Shaper JH (1981). Considerations in the isolation of rat liver nuclear matrix, nuclear envelope, and pore complex lamina. Exp Cell Res 132: 105.

Kernell AM, Ringertz NR (1972). Cytochemical characterization of deoxyribonucleoprotein by UV-microspectrophotometry on heat denatured cell nuclei. Exp Cell Res 72: 240.

Krey AK (1980). Non-intercalative binding to DNA. Prog Mol Subcell Biol 27: 87.

Leemann U, Ruch F (1972). Cytofluorometric determination of basic and total proteins with sulfaflavine. J Histochem Cytochem 20: 659.

LePecq JB, Paoletti C (1967). A fluorescent complex between ethidium bromide and nucleic acids. Physical-chemical characterization. J Mol Biol 27: 87.

Lief RC, Easter HN, Warters RL, Thomas RA, Dunlap LA, Austin MF (1971). Centrifugal cytology. I. A quantitative technique for the preparation of glutaraldehyde-fixed cells for the light and scanning electron microscope. J Histochem Cytochem 19: 203.

MacInnes JW, Uretz RB (1966). Organization of DNA in dipteran polytene chromosomes as indicated by polarized fluorescence microscopy. Science 151: 689.

MacInnes JW, Uretz RB (1967). Thermal depolarization of fluorescence from polytene chromosomes stained with acridine orange. J. Cell Biol. 33: 597.

Malinin G (1978). Permanent fluorescent staining of nucleic acids in isolated cells. J Histochem Cytochem 26: 1018.

Marks GE (1983). Feulgen banding of heterochromatin in plant chromosomes. J Cell Sci 62: 171.

Mazzini G, Giordano P, Riccardi A, Montecucco CM (1983). A flow cytometric study of the propidium iodide staining kinetics of human leukocytes and its relationship with chromatin structure. Cytometry 3: 443. Mello MLS (1983). Cytochemical properties of euchromatin and heterochromatin. Histochem J 15: 739.

Moritz KB (1977). Die Chromosomen von Ascaris in der Keimbahn und in embryonalen Soma. Verh Deutsch Zool Ges 1977: 290.

Moser GC, Meiss HK (1982). Nuclear fluorescence and chromatin condensation of mammalian cells during the cell cycle with special reference to the G_1 phase. In: Genetic Expression in the Cell Cycle (GM Padilla, KS McCarty, eds.), pp 129-147. Academic Press, New York.

_____, Muller H (1979). Cell cycle dependent changes of chromosomes in mouse fibroblasts. Europ J Cell Biol 19: 116.

_____, _____, Robbins E (1975). Differential nuclear fluorescence during the cell cycle. Exp Cell Res 91: 73.

Moutschen J (1976). Fine structure analysis of chromosomes as revealed by fluorescence analysis. Prog Biophys Mol Biol 31: 39.

Müller W, Crothers DM, Waring MJ (1973). A nonintercalating proflavine derivative. Europ J Biochem 39: 223.
Nicolini C, Belmont A, Parodi S, Lessin S, Abraham S (1979). Mass action and acridine orange staining: static and flow cytofluorometry. J Histochem Cytochem 27: 102.
_____, Trefilette V, Cavazza B, Cuniberti C, Patrone E, Carlo P, Brambilla G (1983). Quaternary and quintenary structures of native chromatin DNA in liver nuclei: differential scanning calorimetry. Science 219: 176.
Noeske K (1973). Discrepancies between cytophotometric alkaline fast green measurements and nuclear histone protein content. Histochem J 5: 303.
Pearse AGE (1968). Histochemistry, Theoretical and Applied, 3 rd ed, Vol 1. Little, Brown & Co., Boston.
Popov D, Thorell B (1982). Design and synthesis of new reactive fluorescent dyes for cytofluorometry. Stain Tech 57: 143.
Rasch EM (1984). DNA "standards" and the range of accurate DNA estimates by Feulgen absorption microspectrophotometry. Advances in Microscopy.
Ringertz NR (1969). Cytochemical properties of nuclear proteins and deoxyribonucleoprotein complexes in relation to nuclear function. In: Handbook of Molecular Cytology (A Lima-De-Faria, ed), pp 656-684. Elsevier, New York.
Romen W, Ruter A, Saito K, Harms H, Aus HM (1980). Relationship of ploidy and chromatin condensation in rat liver, moreover a comparison of the nuclear texture in sections and touch preparations. Histochem 67: 249.
Ruch F (1966). Determination of DNA by microfluorometry. In: Introduction to Quantitative Cytochemistry (GL Wied, ed) Vol 1, pp 281-294. Academic Press, New York.
_____, (1970). Principles and some applications of cytofluorometry. In: Introduction to Quantitative Cytochemistry (GL Wied, ed) Vol 2, pp 431-450. Academic Press, New York.
_____, Leemann U (1973). Cytofluorometry. In: Micromethods in Molecular Biology (V Neuhoff, ed), pp 329-346. Springer Verlag, Heidelberg.
Schmitz KS, Ramanathan B (1980). Generation of third-order folded structure for chromatin. J Theoret Biol 83: 297.

Schwartz G, Wittekind D (1982). Selected aminoacridines as fluorescent probes in cytochemistry in general and the detection of cancer cells in particular. Analyt Quant Cytol 4: 44.

Smetana K, Busch H (1966). Studies on staining and localization of acidic nuclear proteins in the Walker 256 carcinoma. Cancer Res 26: 331.

Stockert JC (1972). Meiotic association of X and Y mouse chromosomes as revealed by acridine orange fluorescence after DNA denaturation and differential renaturation. Exp Cell Res 74: 279.

Sumner AT (1981). The distribution of quinacrine on chromosomes as determined by X-ray microanalysis. I. Q-bands on CHO chromosomes. Chromosoma (Berl.) 82: 717.

Swift H (1966). The quantitative cytochemistry of RNA. In: Introduction to Quantitative Cytochemistry (GL Wied, ed) Vol 1, pp 355-386. Academic Press, New York.

Thomes JC, Weill G, Daune M (1969). Fluorescence of proflavine-DNA complexes: heterogeneity of binding sites. Biopolymers 8: 647.

Waring M (1970). Variation of the supercoils in closed circular DNA by binding of antibiotics and drugs: evidence for molecular models involving intercalation. J Mol Biol 54: 247.

Weisblum B (1973). Fluorescent probes of chromosomal DNA structure: three classes of acridines. Cold Spring Harbor Symp Quant Biol 38: 441.

West SS (1969). Fluorescence microspectrophotometry of supravitally stained cells. In: Physical Techniques in Biological Research (AW Pollister, ed) Vol 3, Part C, pp 253-321. Academic Press, New York.

_____, Lorincz AE (1973). Fluorescent molecular probes in fluorescence microspectrophotometry and microspectropolarimetry. In: Fluorescence Techniques in Cell Biology (AA Thaer, M Sernetz, eds), pp 395-407. Springer Verlag, New York.

Wray W, Conn PM, Wray VP (1977). Isolation of nuclei using hexylene glycol. In: Methods in Cell Biology (G Stein, J Stein, LJ Kleinsmith, eds) Vol 16. Academic Press, New York.

Zietz S, Belmont A, Nicolini C (1983). Differential scattering of circularly polarized light as a unique probe of polynucleosome superstructures. A simulation by multiple scattering of dipoles. Cell Biop 5: 163.

Flow Systems:
An Editorial Introduction

Ronald R. Cowden

Most of the light microscopic measurement methods presented in this volume should fall under the general designation of "static systems"; that is, cells or optical objects are viewed and evaluated one-at-a-time. The resulting data can be stored and subsequently reduced to some useful format. As some contributors have pointed out, this approach has the advantage of allowing the investigator to select the material to be measured on the basis of morphological criteria. This is essential when the cell type of special interest is submerged in larger sub-populations of other kinds of cells; or in cases where measurement of nuclear or cell size or shape are important aspects of the experimental design. Flow cytometry systems direct a stream of cells that usually have been fluorochromed past the window of a transducer cell where they are excited and illuminated by light, and the fluorescent emissions of each object are measured either simultaneously or consecutively, along with electronic cell volume or various angles of scattered light. The electronic cell volume (Coulter volume) is proportional to cell size, and both size and nuclear texture contribute to the scattered light signal. The data for each sample is stored or accumulated into electronics essentially similar to those used in scintillation counters (multichannel analysers).

Probably the major special capacity of flow cytometers is that these instruments- even in single-parameter versions-allow sampling of relatively large populations of cells, nuclei or other objects in a relatively short span of time, usually 3-5 minutes per sample. The numbers of cells, nuclei or other optical objects included in each sample allows a form of statistical "overkill" with typical sample sizes of 30,000-50,000, and C.V.'s for DNA measurements are routinely obtained on the order of 2-3 percent. Depending on the system or dye selected, these values can be improved to less than 1 percent.

Flow cytometers have undergone any number of special modifications to adapt them for special purposes: for cell sorting or other experiments with living cells; for higher resolution measurements of compartments in cells or other objects, for measurement of extended objects that must be oriented (chromosomes, sperm); for measurement of polarized fluorescent; for measurement of very small objects; and some attempt has been made to rectify the main failing of flow cytometers-the morphological visualization of the cells measured. With their ability to rapidly process relatively large samples, flow cytometers are almost ideal instruments for studies of cell cycle kinetics or of other factors for which kinetic measurements are essential.

As with commercial image-analysis systems, the principal impetus for the development of these systems has been their potential for use in clinical laboratory medicine. With appropriate fluorochromes, and properly designed instruments using appropriate software, flow systems can be used with success in many cancer diagnosis applications. However, their routine use in laboratory medicine has been for either cell cycle kinetic studies of samples from oncology patients treated by chemotherapy or for the determination and sorting of various subpopulations of cells that can be demonstrated by labeled antibodies. More applications will doubtless follow, and they have become extremely important in research in cell biology and immunology.

Flow Systems:
THE AMAC IIIS TRANSDUCER

Robert C. Leif, Ph.D.
Principal Scientist
Applied Research Department
Coulter Electronics, Inc.
690 W. 20th Street
Hialeah, Florida 33010

A new series of electro-optical flow transducers (AMAC IIIS) have been fabricated. The spherical exterior permits light rays to emerge undeviated and thus effectively lowers the required numerical aperture of the collecting lens. This results in increased light gathering and depth of focus yet permits the optical volume observed to be minimized.

Key terms: AMAC, Electro-optical, transducer, Coulter, fluorescence light scattering, flow analysis, automation and cell.

The information content observable in a single cell to a microscopist is limited by the number of absorbent or fluorescent stains that can be individually discriminated. This information is of course greatly augmented by the morphological information present. However, in terms of automation, this morphological information is ambivalent. In most cases it is sufficient to subjectively describe the cell. Unfortunately this relevant information is often hidden in a very large amount of extraneous data. In contradistinction, flow analysis presently provides very little but extremely precise, specific objective data, which is often sufficient to uniquely describe the cell. It should be noted that the light scattered by cells in flow contains in principle considerable morphological information; however, at present there is no instrument that can make use of all of this data. Future work based on the pioneering studies of the late Paul Mullaney (Burger et al, 1982) may change this situation.

The information provided by a flow transducer to uniquely identify specific cell classes initially increases with the number of independent parameters measured. This realization (Leif 1970) lead to the development of a whole series of multiparameter electro-optical transducers to simultaneously measure electronic impedance changes (Coulter 1953 and Coulter and Hogg 1970) as well as fluorescence and light scattering. The previous AMAC (automated multiparameter analyzer for cells) transducers (Leif et al 1977 and Thomas et al 1977) established the desirability of simultaneous measurement of all parameters as opposed to serial measurements. Besides the obvious advantage of simplicity and consequent economy for simultaneous electro-optical measurements, there is also the advantage of an inherently higher counting rate. Serial measurements require the use of gated electronic windows, which must be wider than the actual transducer output pulse, in order to compensate for any variation of position or velocity between measuring stations. The maximum allowable counting rate of a flow tansducer is inversely related to the square of the pulse width (Steen 1980).

As has been previously reported by Pinkel and Steen (1982), the coefficient of variation (CV) of the optical signal of flow transducer is equal to the square root of the sum of the squares of four individual CV's. The first is associated with the particles. The second with instrument factors unrelated to signal intensity. The third related to photon statistics and the fourth related to background noise, primarily that from light impinging on the photodetector. The contribution of the CV term related to photon statistics until saturation is, as for all microscopes, inversely proportional to the square root of the photon flux. Thus, since the photon flux is proportional to the square of the numerical aperture, the CV is reciprocally related to the numerical aperture. The CV term related to the background photon shot noise is proportional to the square root of the background optical flux, which in turn is proportional to the optical flux, which in turn is proportional to the optical volume.

The AMACIIIS (sphere) transducer (Leif 1982) (Fig. 1) maximizes the numerical aperture while minimizing the imaged optical volume. As previously, ((Leif et al. 1977 and Thomas et al. 1977), the flow chamber is of square cross-section to minimize optical aberration at the aqueous solution quartz interface. However, the outside windows instead of being square are now formed into a sphere. The fluorescent or scattered light resulting from the optical excitation of the cell crosses the transducer-air interface undeviated.

In the case of a square cross-section flow cell with the particle located at its center, the maximum collecting half-angle is 45 degrees or in aqueous solution a numerical aperture of 0.942. Location of the object closer to the flow cell wall will increase the collection angle and consequently the numerical aperture but may interfere with the 90 degree (orthogonal) light scattering.

The refraction of the 45 degree light by the quartz flow cell wall results in a half-angle of 40.31 degrees. This light exits undeviated perpendicularly to the surface of the sphere. The numerical aperture of the collecting lens with a half-angle of 40.31 degrees is 0.647, which even for a long working distance lens is feasible. In the case of the previous square outside window AMACIII transducer, the refraction of the light exiting the quartz into the air resulted in a half-angle of 70.39 degrees or the original numerical aperture of 0.942. The fabrication of a long working distance lens with this large a numerical aperture is not feasible and the depth of focus were it to exist would be less (Piller 1977) than that of the optical system described above for the AMACIIIS, which is a true water immersion optical system.

Two monolithic quartz spherical transducers have been fabricated. The first (Fig. 2) as a sorter tip_R similar in construction to the present Coulter EPICS CVA (cell volume analyzer), which has an identical square inner cross section. The spherical version of the CVA has been demonstrated as had the previous AMACIII and the CVA to serve as a focused flow Coulter Electronic cell volume transducer. The CV of the volume distribution of 5 micron Coulter polystyrene spheres was 2.8%.

Preliminary studies indicate that the flow stream is exceedingly stable as shown by the timed exposure of a fluorescing propidium iodide stream (Fig.3). In a preliminary study a spherical transducer was mounted on a standard EPICS argon ion laser based cell sorter stand and compared with the standard stream in air optics. Propidium iodide labelled Raji cell nuclei were employed as the test particles. Both the spherical transducer and the stream in air performed equally in terms of CV 2.4% (Figs. 4 a and b) with these very bright test particles. In order to avoid saturation of the fluorescence photomultiplier by the horizontal line produced by the interaction of the laser with the stream in air, an obscuration bar has been placed in front of the collecting lens which is perpendicular to the laser beam. Thus the ninety degree obscuration bar could be removed when the sphere transducer was used. This by itself doubled the signal and insignificantly increased the CV to 2.5%. The present EPICS optics are optimized for measurements on a stream in air and thus, although very stable, do not minimize the optical volume viewed. Presently new optics are being constructed which will employ the sphere transducer as an integral element of an optical train which simultaneously minimizes the optical volume and maximizes the numerical aperture.

Acknowledgments

V. De Maria constructed the quartz spherical elements. I would also like to thank N. Files and her staff for staining the cells, R. Feinberg for photography, the Applied Research staff and the EPICS division of Coulter Electronics for providing assistance. I wish to thank G.R. Hibnick, G.A. Liedholz, S.B. Leif and J.C.S. Wood Ph.D. for their reading of this manuscript and the Coulter Corporation for support for this work.

Flow Systems/The AMAC IIIs Transducer / 217

Figure 1: Pair of ray traces of aqueous solution filled quartz square orifice in the AMACIIIS outside spherical surface transducer (left) and the present outside square window transducer (right). The exiting ray is perpendicular to the spherical surface (left) and inclined at 70.39 degrees to the normal (right).

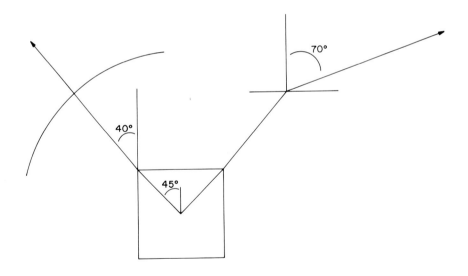

Figure 2: Photograph of a prototype spherical transducer mounted on an EPICS optical assembly.

Figure 3: Photograph of a propidium iodide dye stream in the AMACIIIS transducer.

Figure 4: Flow fluorescence histograms of Raji cells stained with propidium iodide. The ordinate is proportional to the cell number; the abscissa proportionate to the Fluorescence intensity.

(a) Top, conventional stream in air optics with the ninety degree obscuration bar in place (down).

(b) Middle, sphere translucer with the ninety degree obscuration bar in place (down).

(c) Bottom, sphere translucer with the ninety degree obscuration bar raised (up) and thus removed from the optical path; the amplifier gain was halved for this run.

See following page for Figure 4.

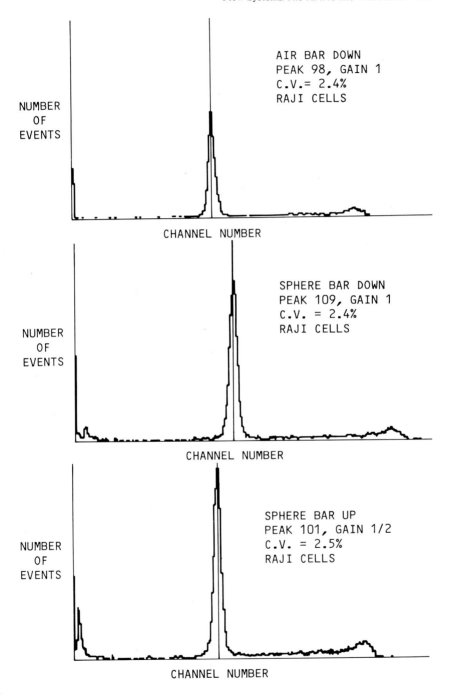

Literature Cited

Burger DE, Jett JH and Mullaney PF: Extraction of Morphological Features from Biological Models and Cells by Fourier Analysis of Static Light Scatter Measurements. Cytometry 2: 327-336, 1982.

Coulter WH: Mans for Counting Particles Suspended in a Fluid. United States Patent #2,656,508, Oct. 1953.

Coulter WH and Hogg W: Signal Modulated Apparatus for Generating and Detecting Resistive and Reactive Changes in a Modulated Current Path for Particle Classification and Analysis (multiple frequency). United States Patent #3,502,974, Mar. 1970.

Leif RC: A Proposal for an Automated Multiparameter Analyzer for Cells (AMAC), Automated Cell Identification and Sorting. Edited by G.L. Wied and G.F. Bahr. Academic Press, New York: 131-159, 1970.

Leif RC: Orifice "Inside Optical Element". United States Patent 4,348,107. 1982.

Leif RC, Thomas RA, Yopp TA, Watson BD, Guarino VR, Hindman DHK, Lefkove N and Vallarino LM: Development of Instrumentation and Fluorochromes for Automated Multiparameter Analysis of Cells. Clin. Chem 23: 1492-1498, 1977.

Piller H: Microscope Photometry. Springer-Verlag 1977.

Pinkel D and Steen HB: Simple Methods to Determine and Compare the Sensitivity of Flow Cytometers. Cytometry 3: 220-224, 1982.

Steen HB: Further Developments of a Microscope-Based Flow Cytometer: Light Scatter Detection and Excitation Intensity Compensation. Cytometry 1: 26-31, 1980.

Thomas RA, Yopp TA, Watson BD, Hindman DHK, Cameron BF, Leif SB, Leif RC, Roque L and Britt W: Combined Optical and Electronic Analysis of Cells with the AMAC Transducers. J. Histochem. Cytochem. 25: 827-835, 1977.

ELECTRON PROBE X-RAY MICROANALYSIS STUDIES ON THE IONIC ENVIRONMENT OF NUCLEI AND THE MAINTENANCE OF CHROMATIN STRUCTURE

Ivan L. Cameron

Cellular and Structural Biology Department
The University of Texas Health
Science Center at San Antonio
San Antonio, Texas 78284

Our working hypothesis is that the ionic environment of chromatin in the nucleus of a living cell plays a major role which determines both the structural and functional organization of the chromatin.

This hypothesis derives support from several sources. For example observations on the chromatin in nuclei which were isolated in the presence of a low ionic strength medium show the nuclei to maintain the in vivo morphology of their interchromatin spaces and to maintain the 25-35nm diameter chromatin fibers (Cameron et al. 1979). When such nuclei were treated with a 140mM solution of monovalent cations (NaCl and/or KCl) several types of nuclear proteins were extracted causing a collapse of the interchromatin spaces but the diameter of the unclumped 25-35nm chromatin fiber was still maintained. When such nuclei were washed in distilled water, the nuclei swelled and the 25-35nm chromatin fiber unfolded into a beaded string of nucleosomes 10nm in diameter. This unfolded chromatin was still sensitive to the ionic environment in that the higher ordered 25-35nm structure of chromatin again returned when the NaCl concentration in the environment was raised to 10mM or higher (Brasch et al. 1971, Cameron et al. 1979, Murcia and Koller 1981, Hancock and Boulikas 1982).

Also, Kellermayer and Hazlewood (1979) using a staining polarization optical technique showed differences in the structural organization of DNA of chromatin in interphase nuclei and in mitotic chromosomes. Mitotic chromatin was birefringent while interphase chromatin was non-birefringent but was made birefringent by exposure to solutions of Na^+ and/or K^+ at concentrations of 120mM.

Thus, it seems possible that the chemical activities of monovalent ions in the nucleus of living cells may play a major role in the structural organization of interphase and mitotic chromatin in vivo. From the above in vitro observations one may speculate that the chemical activity of the monovalent cations in the environment of the interphase chromatin is maintained at less than 120mM, but higher than 10-15mM. These in vitro observations also suggest that a chemical activity of 120mM or greater may result in the removal of certain chromatin associated proteins and in the condensation of chromatin into mitotic or meiotic chromosomes.

In fact, recently Kellermayer et al. (1984) have presented data to support the idea that lower chemical activities of monovalent cations stabilize the chromatin-solute-water interactions when the concentration of Ca^{2+} and Mg^{2+} ions are held constant. They also speculate that any normal or pathological conditions that might cause the total chemical activity of monovalent cations to increase significantly above 30mM will destabilize chromatin structure in the interphase nucleus of living cells.

Thus, these studies on isolated nuclei suggest that the chemical activity of total monovalent ions (Na^+ plus K^+) in the interphase nucleus of living cells probably ranges between a low of 10mM and high of 120mM. These studies also suggest that chromatin in the living cell may experience a rise to higher concentrations at the time of chromatin condensation for meiosis or mitosis.

This led us to ask what type of ionic environment is chromatin actually exposed to within the living cell? To help answer this question we turned to electron probe X-ray microanalysis, which provides the capability of accurately measuring the concentration of a number of elements at the subcellular level in cells which are appropriately prepared using cryofixation, cryosectioning and cryoadsorption procedures. However, before proceeding it is important to point out that the determination of ion concentration on a dry weight or even on a wet weight basis in a subcellular compartment is not necessarily a measure of the chemical activity of that ion. This is because the measured chemical activity of an ion can be different from the ionic concentration in the cells if the ion is compartmentalized, sequestered or adsorbed. The classic example of this is in the case of cellular Ca^{2+} where the cellular concentration as measured by chemical procedures is in the mM range whereas the measured chemical activity is in the nM to μM range.

Because the intranuclear chemical activity of monovalent cations is what we really need to know, we must either resort to ion-selective electrodes or other direct method of measuring the chemical activity of ions or we must turn to less direct approaches for determination of the chemical activity of the ions in the nucleus.

<u>Validation of our electron-probe X-ray microanalysis methods by obtaining quantitative agreement with the known and accepted wet chemical measurements.</u>

For electron probe x-ray microanalysis fully grown amphibian oocytes of <u>Rana pipiens</u> and of <u>Xenopus laevis</u> were placed on a brass pin (Cameron et al., 1983b). The specimens were then frozen by immersion in liquid propane cooled in a liquid nitrogen bath and were then transferred to and stored in liquid nitrogen until the time of sectioning. Sectioning was done on the LKB Ultratome V equipped with a modified cryokit and cooled to a temperature of either -40°C or -80 to -100°C for both specimen and knife. A dry glass knife with a 40° angle and a sectioning speed of 0.5mm/s was used for cutting. Ultrathin sections were positioned on a film of formvar (0.25% in dioxane) spanning a 1.5mm hole in a 35mm carbon grid. To minimize curling or movement of the sections, a carbon-coated formvar film on an aluminum ring was placed over the sections. The sandwiched specimen was dried within the LKB chamber at -40°C or -100°C in a custom-made cryosorption apparatus by evacuation with a rotary pump for 2 hr. The sections were warmed to room temperature, vented with dry nitrogen gas, and stored in a desiccator. At the time of analysis the aluminum ring was removed, leaving a flat section sandwiched between two layers of formvar film. The sections were examined in a JEOL JSM-35 scanning electron microscope under the following conditions: STEM mode, accelerating voltage 25kV, specimen-to-detector distance 15mm. Analysis was done by a Si(Li) x-ray detector and Tracor Northern NS-880 x-ray analysis system.

Our quantification technique is based on the Hall mass fraction method (1971, 1973). Continuum counts, due to the formvar, were measured and were subtracted prior to the calculation of elemental peak-to-continuum values, which were converted to content by a series of cryosectioned standards with known amounts of dried salts added to a 20% bovine serum albumin solution.

Analysis of variance statistics was used throughout. Where a significant F value was obtained, a Student-Newman-Keul's multiple range test was used to determine significant differences between the means of each treatment procedure.

A cryomicrodissection technique has been developed, which prevents artefactual solute distribution and permits the isolation of nuclei from large cells such as amphibian oocytes, which allows quantitative comparison to our microprobe procedures (Century and Horowitz, 1974). The isolated nuclei are collected, weighed, freeze-dried, reweighed, and subjected to chemical analysis. The nuclear concentrations of Na^+ and K^+ in the nuclei of fully grown oocytes from the leopard frog are listed in Table 1. We in our laboratory have recently performed microprobe analysis of freeze-dried sections of mature oocytes of the leopard frog. Assuming an 85% water content, as measured and reported by Century and Horowitz (1974), we have converted our dry weight values to wet weight values for comparative purposes in Table 1. Data values for Na^+ and K^+ from the two different cryopreparative techniques are in striking agreement. This is indeed encouraging evidence that microprobe can be used to make accurate measurements at the subcellular level.

Table 1. Values for Na^+ and K^+ concentrations (mM/l water) of nuclei of mature oocytes of <u>Rana pipiens</u> (leopard frog)

Technique	Na^+	K^+
Ultralow temperature microdissection plus weighing and flame photometry[a]	7.3	126
Microprobe[b]	7.9	115

[a] Century and Horowitz (1974) (n=7-10 nuclei).

[b] LaBadie et al. (1981) and assuming 85% nuclear water as measured by Century and Horowitz (1974) (n=13 measurements on 3 oocytes).

An additional experiment was run to test the effect of the cryosectioning and cryosorption temperature and the extracellular ionic environment on the measured concentration of Na^+ and K^+ in yolk-free cytoplasm measured near the nucleus of large fully grown oocytes of <u>Xenopus laevis</u>. The results which are shown in Table 2 indicate that temperatures of $-40°$ or lower are adequate to prevent long-range diffusion of Na^+ and K^+ in these large oocytes.

Table 2. Effect of cryosectioning and cryosorption temperature and extracellular ionic environment on the concentration of sodium and potassium in the yolk-free cytoplasm adjacent to the nucleus in large fully grown oocytes of <u>Xenopus laevis</u>[a] (mM/kg dry weight, mean ± SE[b])

Temperature and Environment	Sodium	Potassium
$-80°C$, Ringer's	60.7±14	428±73
$-40°C$, Ringer's	53.7±14	436±42
$-40°C$ Mineral Oil	57.5±7	354.37

Statistics of 1-way ANOVA

F Value	0.060	0.601
p Value	NS	NS

a Vitilline membrane manually removed from each oocyte

b mean of three oocytes with 6-8 independent areas microprobed per oocyte

From this analysis we feel safe in concluding that we have developed and validated a procedure for the accurate quantification of intraoocyte electrolytes using energy-dispersive electron-probe x-ray microanalysis of thin cryosections.

Notice that the total concentration of the monovalent cations (Na^+ plus K^+) in the nucleus of the oocyte, when converted to a wet weight basis, as reported in Table 1 is 123mM. This is high enough that it might be expected to cause extraction of proteins plus structural change in the chromatin of isolated nuclei as reported above. We therefore asked ourselves if the ionic concentrations as reported in Table 1 were a true measure of free ionic strength in the nucleus? The next section of this report gives a brief account of our attempt to answer this important question.

Use of electron probe X-ray microanalysis procedures to help establish the extent of the free and the bound state of K^+ in the nucleus of amphibian oocytes.

Study of the size and kinetics of slower and faster ion exchange fractions in whole cells or in subcellular fractions has proved difficult because of the numerous assumptions and unknowns involved in such studies. Most commonly such studies involve the exposure of a population of cells to a medium of known ionic composition often containing a radioactive isotope of an ion of interest. Analysis of the water and ion(s) of interest is then done in cell-pellets packed at a given g force. The cells are often washed during separation from the incubation medium and the trapped extracellular space in the cell pellets is measured independently using a trace amount of a radioactive extracellular marker substance.

Although this procedure has been the most commonly used approach to study ion exchange in cells it is subject to methodological questions and to questions of data interpretations (Negendank and Shaller, 1980). For example, one may ask: what is the contribution from the extracellular medium or the wash procedure, what is the contribution from the surface matrix of the cells, and are subpopulations of deteriorating or other cell types present?

Interpretation of the data involves another series of questions concerning the contributions of various factors such as: adsorption or desorption from intracellular macromolecules, plasma membrane permeability properties, and different exchange fractions in the subcellular compartments. The introduction of quantitative energy dispersive electron probe x-ray microanalysis (EDS) of electrolytes at a subcellular level in thin cryosections allows us to circumvent the above methodological questions and to be much more precise in our interpretation of the data. In this experiment we subjected manually defolliculated fully grown oocytes from the amphibian <u>Xenopus laevis</u> to Ringer's solution in which rubidium (Rb^+) was substituted as a surrogate for potassium (K^+) on a molar basis. Samples of oocytes were serially removed from the Rb^+-containing Ringer's solution during a 24 hour period and processed for EDS to measure K^+ efflux. The subcellular concentration of Na^+ and Cl^- was also measured during the experiment. Figure 1 summarizes the data on K^+ efflux from the nucleus of the oocytes.

Briefly, the data from this study (Cameron and Hunter, 1984) indicated: that K^+ efflux had two exchange components in the nucleus, that there was a faster and slower flux fraction, and that the faster efflux fraction accounted for 14% of the total K^+ present, and that Rb^+ exchanged for K^+ but not for Na^+.

Based on our interpretation of these data, the fast K^+ exchange fraction represents the free K^+ fraction and therefore only about 14% of the intranuclear K^+ is free to contribute to the chemical activity of this ion in the nuclear environment. If we take 14% of the measured K^+ concentration in the nucleus from table 1 and we assume that all of the intranuclear Na^+ is free in solution we can calculate that the total concentration or chemical activity of these two monovalent cations combined is 26mM. This is our best estimate of the chemical activity of monovalent Na^+ + K^+ ions in the interphase nucleus of a large oocyte, the chromatin of which is considered to be in a dispersed state. This is within the chemical activity range where the chromatin in isolated nuclei was maintained in a stable structural state as discussed above.

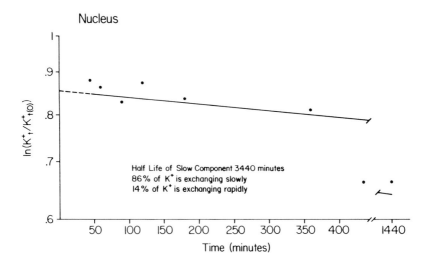

Fig. 1. Results of K^+ efflux studies in the nuclear compartment. Oocytes were placed in Ringer's solution where Rb^+ was substituted for K^+. Samples were serially removed for ion analysis over a 1400 min incubation period. Data were analyzed by extrapolating back to the intercept. This slow K^+ exchange fraction accounts for 86% of the total K^+ and the fast K^+ exchange fraction accounts for 14% of the total K^+.

A basic premise of our calculations so far is that the water is totally free to act as a solvent within the oocyte nucleus. The following section addresses this question from an indirect and preliminary set of observations, which relates the size of ice crystals to water content and physical state of water in subcellular compartments.

Use of ice crystal size as an indicator of water content and the structural state of water in the nucleus and in the cytoplasm.

This study was aimed at obtaining information on the content and on the physical state of water as compared to bulk water in subcellular compartments and is based on a recently completed study on this subject (Cameron et al., 1984). The approach involves a correlative analysis of ice crystal size, water proton NMR relaxation times and water content data obtained from measurements on a number of different biological materials including macromolecular solutions, extracellular fluids and materials, tissue cell preparations and intracellular compartments.

The rationale was to see if the local size of ice crystals, produced under controlled freezing conditions, would give information that could be related to the water content and to the mobility of the water as determined from NMR relaxation time measurements. It was our hope that local ice crystal size alone would prove a useful source of information about the water content and the physical state of water at the subcellular level. In this way we might begin to learn more about water in cells.

In brief we found that larger ice crystal size is significantly correlated to higher water content and to longer T_1 and T_2 relaxation times. Thus, in the majority of cells and tissues (i.e., those which contain primarily globular proteins) ice crystal size was a significant predictor of water content and of T_1 and T_2 relaxation times. It therefore appears that the size of ice crystals can indeed give information on the amount and on the physical state of water at the subcellular level (Table 3). In this regard, an observed larger size of ice crystals in the nucleus as compared to the open cytoplasm in several cell types including the oocyte, fig. 2, indicated that the nucleus is generally higher in water content and has longer T_1 and T_2 relaxation times than does open cytoplasm.

Table 3. Calculated water content and water proton NMR relaxation time values in cellular compartments based on correlations to measured ice crystal size[a]

Tissue and subcellular compartment	Measured ice crystal size area (μm^2)	Percent water calculated from correlation with ice crystal size	T_1 and T_2 relaxation times (msec) calculated from correlation with ice crystal size	
			T_1	T_2
Xenopus oocyte (full grown)				
nucleus	0.164	98.9	1733	575
nucleolus	0.031	68.2	256	100
yolk-free cytoplasm	0.040	70.2	356	132
yolk	0.000	---	---	---
erythrocytes	0.043	70.9	384	143
actual measured values		71	461	109
plasma	0.104	85.0	1067	361
actual measured values		91	1164	312

[a]Data from Cameron et al. (1984).

In this regard, the water in many cells and subcellular compartments does not have the freezing characteristics of dilute aqueous solutions; however, the water in the nucleus of the fully grown amphibian oocyte did appear to be more like bulk water than was the water found in other subcellular compartment or other cell types, Table 3. Thus an assumption that most of the water in the oocyte nucleus is like bulk water seems justified in this case.

One observation from this study raised an interesting question. How can immediately adjacent areas such as the nucleoplasm and the nucleolus which are not separated by a membrane have such vastly different ice crystal sizes (Table 3 and figure 2a). If ice crystal size is an accurate indicator of the percent water content then one might think that the water in the open nucleoplasm would be free to move into the nucleolus to balance or to bring to equilibrium the concentration of water in both compartments. This is of course not what was observed. The following is offered as one possible explanation of the observation. The nucleolus of Xenopus laevis is known to have a tightly wound core of

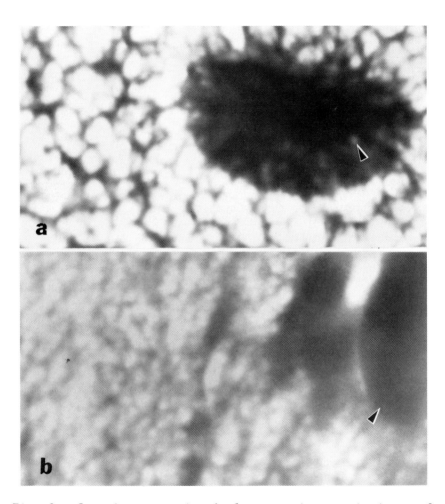

Fig. 2. Scanning transmitted electron microscopic image of a 2μm thick freeze-dried cryosection of an oocyte showing ice crystal reticulations in the nuclear area, fig. 2a, and in the cytoplasm area, fig. 2b (13,500X). The nucleoplasm has larger crystal size than the cytoplasm. The nucleolus has smaller ice crystals (arrow) than the nucleoplasm and the yolk platelet (arrow) shows no ice crystal formation.

long coarsely coated filament (Miller and Beatty, 1969). The filament is a single DNA molecule irregularly coated with ribonucleo-protein (RNP) matrix (Miller, 1966). When the nucleolus is isolated and dispersed the RNP matrix is seen to consist of numerous individual fibers. We propose that the abundant filamentous network of the nucleolus in situ orders a large fraction of the nucleolar water into a hydration shell and that the tightly wound core of filamentous material provides a physical restraint which resists the intrusion of less ordered nucleoplasmic water into the interstices of the nucleolus.

So far we have dealt with the ionic environment of the interphase nucleus. We now ask ourselves if physiological changes in the ionic environment of the nucleus can be correlated to structural and functional changes in chromatin in the living cell.

Changes in the intranuclear ionic environment during meiotic maturation of Xenopus oocytes.

Fully grown amphibian oocytes arrest at a late stage of first meiosis. Such arrested oocytes can be removed from the ovaries and can be induced by progesterone to undergo maturation which consists of: germinal vesicle (nuclear envelope) breakdown (GVBD), condensation of the chromatin into chromosomes, completion of the meiotic division and the extrusion of the first polar body. After completion of the first meiosis the oocyte normally becomes an egg and is capable of being activated or fertilized to undergo the second meiosis. Here we can ask if the condensation of chromatin into meiotic chromosomes is accompanied by an increase in the intranuclear concentration of monovalent cations, specifically Na^+ plus K^+.

To determine the role of ionic events in this process, we, and others, have measured the time course of changes in oocyte: volume, water content, concentration of Na^+, K^+, Cl^-, Ca^{2+}, Mg^{2+} and H^+ and the NMR spin-lattice (T_1) relaxation time of oocyte water protons (see figure 3).

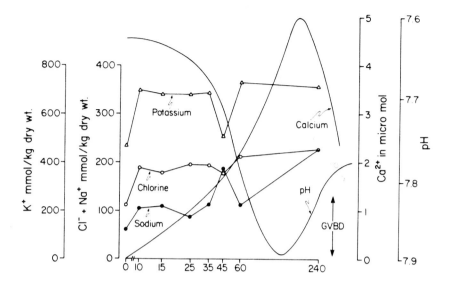

Figure 3. Summary of changes in the concentration of ions during maturation of <u>Xenopus</u> oocytes. The Na^+, K^+ and Cl^- data are taken from an electron probe X-ray microanalysis study (Cameron et al. 1983a, LaBadie and Cameron, 1981), the hydrogen ion data are taken from a pH microelectrode study (Lee and Steinhardt, 1981), while the Ca^{2+} data are taken from a calcium ion selective microelectrode study (Moreau et al. 1980).

The influence of amiloride, a drug reported to inhibit passive influx of Na^+ into stimulated cells, on both GVBD and Na^+ content was also measured. The results are compatible with the following hypothesis. Initially the oocyte has a low intracellular free Na^+ and free K^+ concentration. Progesterone stimulation causes an increase in the motional freedom of water molecules and increases the permeability of the oocyte membrane to Na^+ and K^+ influx but also promotes their release from bound intracellular stores. Some of these stores are closely associated with macromolecules. An increasing intracellular free Na^+ may enhance a Na^+/Ca^{2+} exchange at available Ca^{2+} storage sites, such as mitochondria, cytoplasmic vesicles, calcium binding macromolecules, pigment granules or cortical granules. Which of these sites that will be drawn upon depends on which of the sites contain sufficient Ca^{2+} and on the presence of external Ca^{2+}. Once the free Ca^{2+} surge reaches a threshold point maturation events proceed and the free Ca^{2+} surge by itself is sufficient to allow GVBD. It also appears that the amiloride-sensitive surge of free Na^+ at 45 min is temporally coupled to a Na^+/H^+ exchange allowing an alkaline shift. That oocyte maturation can occur in isosmotic sucrose indicates that intracellular ionic stores are adequate to permit necessary free ionic surges and that ionic influx is not essential for meiotic maturation in this species. Finally, it appears that the intranuclear concentration of monovalent cations reaches a peak at about the time of rapid condensation of chromatin into meiotic chromosomes.

The observations of Kellermayer and Hazlewood (1979) which showed that interphase chromatin could be made to stain like mitotic chromosomes by exposing them to solutions of Na^+ and/or K^+ at concentrations of 120mM are in agreement with our observations that the intranuclear concentration of the cations naturally increases at the time of chromosome condensation and GVBD. In fact if we again assume: that about 14% of the concentration of K^+ is free, that all of the Na^+ is free and that the water content of the nucleus is about 85% in the oocyte nucleus, then the sum of these free monovalent cations in the environment of nuclear chromatin would appear to increase two-fold at the time of chromatin condensation just prior to breakdown of the nuclear envelope. It therefore seems possible that the processes of meiosis and of mitosis normally involve a surge in the chemical activity of these monovalent cations which may in turn cause the loss of specific intranuclear proteins which were involved in the stabilization of interphase chromatin.

Summary

We have demonstrated the application of electron probe X-ray microanalysis to studies aimed at determining the role of the ionic environment in the maintenance of a stable chromatin structure during interphase. Our results indicate that X-ray microanalysis can be used to accurately measure the concentration of ions, such as Na^+ and K^+ within subcellular compartments. We also present evidence from our X-ray microanalysis studies to show that only a portion of the K^+ in the interphase nucleus is in a free ionic state and that the rest of the intranuclear K^+ is adsorbed to intranuclear macromolecules. The stable structural state of interphase chromatin was correlated with the maintenance of a normal range in the chemical activities of monovalent ions in the oocyte nucleus. Finally, we demonstrated that the chemical activity of the monovalent ions probably increases beyond this normal interphase range as the chromatin in the nucleus condenses into discrete chromosomes during the first meiotic division in the amphibian oocyte.

The above observations support the hypothesis of Kellermayer et al. (1984) that the maintenance of chromatin structure in the nucleus of living cells requires a balance between the presence or absence of mobile nuclear proteins and the chemical activity of monovalent cations. It also suggests that an elevation of the chemical activity of the monovalent cations above a normal range will result in the destabilization of interphase chromatin structure. It is hypothesized that an elevation of monovalent cations may be directly involved in the structural and functional changes in chromatin which are known to take place at the time of meiosis and mitosis.

REFERENCES

Brasch K, Selegy VL, Setterfield G (1971). Effects of low salt concentration on structural organization and template activity of chromatin in chick erythrocyte nuclei. Exp Cell Res 65:61-72.

Cameron, IL, Pavlat, WA, Jeter, JR Jr. (1979). Chromatin substructure; An electron microscopic study of thin-sectioned chromatin subjected to sequential protein extraction and water swelling procedures. Anat. Rec. 194:547-562.

Cameron IL, LaBadie DRL, Hunter KE, Hazlewood CF (1983a). Changes in water proton relaxation times and in nuclear to cytoplasmic element gradients during meiotic maturation of Xenopus oocytes. J Cell Physiol 116:87-92.

Cameron IL, Hunter KE, Smith NKR (1983b). Validation of quanitative energy-dispersive electron-probe X-ray microanalysis of electrolytes in thin cryosections of erythrocytes. In Gooley R (ed): "Microbeam Analysis-1983", San Francisco:San Francisco Press, Inc, p237-242.

Cameron IL, Hunter KE (submitted 1984). Fast and slow potassium-rubidium exchange fractions in the nucleus and in the cytoplasm of amphibian oocytes.

Cameron IL, Hunter KE, Ord VA, Fullerton GD (submitted 1984). Ice crystal size as an indicator of water content and water proton NMR relaxation times in cells.

Century TJ, Horowitz SB (1974). Sodium exchange in the cytoplasm and nucleus of amphibian oocytes. J Cell Sci 16:465-471.

Hall TA (1971). The microprobe assay of chemical element. In Oster G (ed): "Physical Techniques in Biological Research", Vol IA, New York:Academic Press, pp. 157-159.

Hall TA, Anderson CH, Appelton T (1973). The use of thin specimens for X-ray microanalysis in biology. J Microsc 99:177-182.

Hancock R, Boulikas T (1982). Functional organization in the nucleus. Internatl Review of Cytol 79:165-214.

Kellermayer M, Hazelwood CF (1979). Dynamic inorganic ion-protein interactions in structural organization of DNA of living cell nuclei. Cancer Biochem Biophys 3:181-188.

Kellermayer M, Rouse D, Gyorkey F, Hazlewood CF (1984). Ionic milieu and volume adjustments in detergent-extracted thymic nuclei. Physiol Chem and Phys and Med NMR (in press).

LaBadie DRL, Cameron IL (1981). Nuclear and cytoplasmic changes in sodium content during progesterone induced maturation of Xenopus oocytes: Involvement of net sodium flux. J Cell Biol 91:2a.

LaBadie DRL, Pool TB, Smith NKR, Cameron IL (1981). Element content in the nucleus and in yolk and yolk-free cytoplasm of developing Rana pipiens oocytes: An X-ray microanalysis study. J Cell Physiol 109:91-97.

Lee SC, Steinhardt RA (1981). pH changes associated with meiotic maturation in oocytes of Xenopus laevis. Dev Biol 85:358-369.

Miller OL, Jr. (1966). Structure and composition of peripheral nucleoli of salamander oocytes. Natl Cancer Inst Monograph 23:53-66.

Miller OL, Jr, Beatty BR (1969). Nucleolar structure and function. In Lima-De:Faria A (ed): "Handbook of Molecular Cytology", Amsterdam, North-Holland Publishing Company, Inc, pp. 605-619.

Moreau M, Vilian JP, Guerriu P (1980). Free calcium changes associated with hormone action in amphibian oocytes. Dev Biol 78:201-214.

Murcia G, Koller T (1981). The electron microscopic appearance of soluble rat liver chromatin mounted on different supports. Biol Cell 40:165-174.

Negendank W, Shaller C (1980). Multiple fractions of sodium exchange in human lymphocytes. J Cell Physiol 104:443-459.

Smith NKR, Cameron IL (1981). Observations on electron probe X-ray microanalysis compared to other methods for measuring intracellular elemental concentration. Scanning Electron Microscopy 1981 II:395-408.

ULTRASTRUCTURAL METHODS

ULTRASTRUCTURAL HISTOCHEMISTRY WITH THE HIGH-VOLTAGE ELECTRON MICROSCOPE (HVEM)

Eugene L. Vigil[1], William P. Wergin[2], M. N. Christiansen[2]

[1]Department of Horticulture, University of Maryland, College Park, MD 20742.
[2]Plant Stress Laboratory, PPHI-BARC, Beltsville, MD 20705.

INTRODUCTION

Electron images obtained from biological specimens contain a wealth of information on membrane organization and organelle compartmentalization for eucaryotic cells. These data enabled cell biologists to identify and separate membranous organelles into categories based on size, shape, internal organization and, to some extent, position within the cell. Information on the three-dimensional organization of membranous organelles, however, could only be inferred from two-dimensional images obtained with conventional low voltage electron microscopes. One way to solve this problem is by serial section reconstruction to reproduce a three-dimensional model. This is, however, a labor intensive procedure requiring precision in sectioning, accuracy in handling many serial sections, and often deletion of much information during translation of only a portion of the image in electron micrographs to transparencies used for 3-D reconstruction. These factors have undoubtedly contributed to the limited number of published studies using serial section reconstruction. There still exists, however, a positive interest and unquestionable need for definitive information on the three-dimensional shape and contour of cellular organelles, their physical relationships to each other, and what structural changes these organelles undergo in response to various physiological conditions associated with normal and abnormal growth and development of cells. The development of the high voltage electron microscope (HVEM) made it possible for the first time to observe directly three-dimensional aspects of organelles through examination of sections of biological specimens up to several micrometers thick (see reviews by Dupouy, 1968; Cosslett, 1971; Hama, 1976; Glauert, 1979; Hawes, 1981). A high accelerating voltage in the range of a million electron volts (1,000 kV), which was essential for effective penetration of thick specimens, provided images of cell structures at a level of resolution (ca. 2.5 nm) comparable to that measured in images for sections 30 times thinner (Hama, 1976).

The abundance of cellular information contained in a thick section cannot be visualized adequately in the standard two-dimensional images obtained with HVEM due to superposition of structures in space. An exception would be peripheral regions of migrating cells where filamentous microtrabeculae are visible with little interference from other organelles (Wolosewick and Porter, 1976). Although these images are useful, more specific spatial relationships are available with stereoscopy of tilted specimens (Beeston, 1973; Hudson, 1973; Wolosewick and Porter, 1979). A remaining problem with thick sections, even when examined with stereoscopy, is that fine details of cellular structures are not readily visible unless they have selective affinity for heavy metals, e.g., chromosomes (Ris, 1969), pigment granules (Byers and Porter, 1977). The application of cytochemical procedures to different tissues not only overcomes the problem of organelle contrast but permits selective staining of specific organelles. Coupled with HVEM, cytochemical staining effects visualization of significant new information on organelle organization in sections up to 10 m thick. The use of metal impregnation (osmium alone or zinc iodide-osmium complex) or enzyme activity has made it possible to obtain very elegant images of the cis or forming face of the Golgi apparatus (GA) (Hermo et al., 1979; Rambourg et al., 1979; Favard and Carasso, 1973), endoplasmic reticulum (ER) and nuclear envelope (NE) (Favard and Carasso, 1973; Harris, 1979; Hawes et al., 1981; Poux, 1981), and plant vacuoles (Marty, 1973, 1978, 1980; Poux et al., 1974).

This report presents information on for the three-dimensional organization of several organelles of the

endomembrane system, i.e. plasma membrane (PM), GA, ER and
NE, based on observations of thick sections of cytochemically stained tissue viewed with HVEM. Certain cell
types were chosen for cytochemical staining and HVEM where
a particular organelle has a prominent and major
involvement in cellular function: (1) The uptake of tracer
protein by endocytosis along the luminal PM of muscle
capillaries illustrates dramatic changes in surface
topography; (2) Formation of lysosomes from the GERL
network of cisternae and tubules of the GA in granulosa
cells of regressing copora lutea demonstrates intricate
and complex aspects of lysosome formation; (3)
Organization and polarized distribution of cisternal and
tubular ER in epithelial cells of the proximal tubule
provide clear evidence of the importance of ER in
intra-organellar communication; (4) The processes of
mitosis and cyto-kinesis in the radicle apex, involving
alignment of ER and NE components as part of the spindle
apparatus and phragmoplast-cell plate complex, direct
attention to the importance of cellular organelles other
than microtubules, chromosomes, and GA in the process of
cell division.

SPECIMEN PREPARATION

Preparation of specimens for HVEM follows established
procedures of fixing and embedding tissue for electron
microscopy. This also applies for performing cytochemical
reactions with aldehyde-fixed tissue. We routinely use 20
to 50 μm sections cut with a vibratome or tissue slicer
for cytochemistry: the reactants are more accessible to
all cells in the section and the diffusion gradient is
markedly less than that in small cubes of tissue. Thick
(0.2 to 3.0 μm) sections of embedded tissue are cut
alternately between thin sections (Favard and Carasso,
1973). Sections are usually picked up on formvar-coated
slot grids which provide maximum free area for tilting
($\pm$ 30°) any region of the specimen. The decision to
counter stain sections with uranyl or lead salts (Carasso
et al., 1973) depends on whether or not additional
information on neighboring organelles is desired (compare
Figs. 2 and 3). This applies mainly to sections less than
1 μm in thickness (Favard and Carasso, 1973). All figures
shown in this report are stereo pairs with stereo
labelling. Wolosewick and Porter (1979) provide specific
instructions for viewing stereo pictures without stereo

glasses.

ENDOCYTOSIS

One of the advantages of examining thick sections with HVEM is that organelles containing a small amount of cytochemical reaction product, which may not be visible in thin sections, are readily apparent with HVEM, as in Fig. 1. This unstained section (ca. 0.5 µm) of an endothelial cell in the rat diaphragm muscle reveals distribution of a tracer probe selectively stained by enzyme cytochemistry. Loading of endocytic vesicles (EV) with tracer molecules defines location, size, and density of forming vesicles and the outline of cellular junctions (CJ). For this preparation myoglobin, a small heme protein with weak peroxidase activity, had been used as a tracer molecule to determine the route of transport from the blood to tissue (Anderson, 1972; Anderson and Vigil, 1979; Simionescu et al., 1973). The rat diaphragm was chosen because it is a good test tissue having adequate thinness for rapid penetration of fixative for short time tracer studies, is readily accessible for in situ fixation, and has numerous capillaries. Sites of myoglobin distribution are visualized by a peroxidase reaction using diaminobenzidine as substrate (Graham and Karnovsky, 1966) which is converted to osmium black during osmication of tissue (Hanker, et al. 1964).

Within 60 seconds following i.v. injection of a 10% solution of myoglobin, loading of numerous endocytic vesicles occurred along the luminal PM in addition to filling endothelial junctions (CJ). It is remarkable that almost the entire surface of the PM is involved in endocytosis, the only limitation for endocytosis appears to be available space along the PM. There is, however, as noted by Simionescu et al (1973), a limited endocytic process occurring along the PM in the parajunctional region (PJ). Most of the labeled EV appear still attached to the PM. The short connecting stalks observed by Simionescu et al. (1973) are not readily visible in this stereo pair. Interestingly, stalks may break as vesicles separate and pass through the cytoplasm toward the opposite side of the cell or become part of an open channel or fenestra between the capillary lumen and the surrounding tissue (Simionescu, 1981). Endocytic vesicles clearly represent an important component of endothelial cells, being the morphological equivalent of the physiological large and small pore compartments of capillaries (Simionescu et al., 1975).

Figure 1. This 0.5 m section is through a diaphragm capillary of rat previously injected i.v. with myoglobin solution. Diaphragm fixed _in situ_ within 1 min of injection and processed for peroxidatic activity. Osmium black fills the capillary lumen and endocytic vesicles (EV) along the luminal PM. Fewer EV are present along the PM in the parajunctional area (PJ) or cell junction (CJ) than elsewhere. The weak peroxidatic activity of cytochromes is responsible for staining mitochondria cristae (M). N, Nucleus Section not counter stained with heavy metals. Bar on this and all figures equals 1.0 m. X 12,000; 15° tilt.

Figure 2. Golgi apparatus (GA) in granulosa cell of 15 day corpus luteum of ewe. Tissue reacted for acid phosphatase, being confined to GERL cisternae, associated coated vesicles (CV) and neighboring lysosomes (L). Secretory vesicles (SV) are not reactive for acid phosphatase. ER, endoplasmic reticulum; M, mitochondria. This 0.5 μm section counter stained with uranyl and lead salts. X 16,800; 16° tilt.

LYSOSOME FORMATION - GERL CISTERNAE

Since the discovery of GERL cisternae by Novikoff (1964), there has been considerable dispute regarding origin and function in different cell types (Novikoff, 1976; Novikoff and Novikoff, 1977; Hand and Oliver, 1977; Goldfisher, 1982). The view emerging from these and other studies is that GERL cisternae are integral trans elements of GA, stain positive for acid phosphatase, and are involved in formation of lysosomes and possibly secretory vesicles (see review by Goldfisher, 1982). Of the different cell types where GERL has been studied, the granulosa cells of the corpus luteum represent a good system for application of enzyme cytochemistry and HVEM, because numerous lysosomes appear during normal (McClellan et al., 1977; Paavola, 1978a,b) and prostaglandin-induced (McClellan et al., 1977) cellular regression. These studies, while providing evidence for the formation of

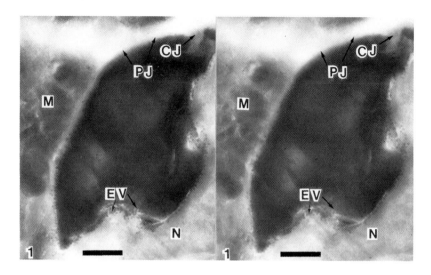

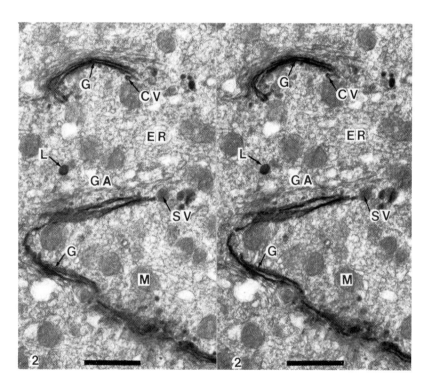

lysosomes in close proximity to GERL cisternae, did not demonstrate an unequivocal relationship for lysosome formation from GERL.

One inherent problem in determining the site of lysosome formation in relation to GERL is the difficulty of obtaining within a single thin section evidence for direct continuities between lysosomes and GERL. This difficulty is due to the shape and cellular space occupied by cisternae and fenestrated tubules of GERL. In granulosa cells, GERL cisternae, as revealed by acid phosphatase activity, consist of central continuous saccules which normally follow the contour of preceding trans elements of the GA (G in Fig. 2) and a peripheral fenestrated tubular (T) network where several lysosomes arise (Figs. 3, 4). Figures 3 and 4 illustrate further the elegant coordination of structural and functional changes in GERL during induced regression. Dense bodies (DB) form from tubules in the fenestrated network; whole cisternae are lost through tubule fusion and actual sloughing to form autophagic vacuoles (AV in Figs. 3, 4). GERL elements are most likely replaced by differentiating trans cisternae immediately behind those removed in the process of forming lysosomes. Lysosome formation appears to be a well coordinated and regulated process in granulosa cells of the ewe corpus luteum; one that can be accelerated by prostaglandin F2 alpha-treatment (McClellan et al., 1977).

ENDOPLASMIC RETICULUM

Three-dimensional aspects of ER in animal and plant cells are often difficult to discern even with selective staining and HVEM (Favard et al., 1973; Poux, 1981). The main reason is that this membrane system, by definition, traverses all areas of the cell outside the nucleus without any major degree of spatial asymmetry. In the following discussion we present two examples illustrating prominent patterns of ER organization: (1) Columnar cells of the proximal tubule of nephrons; (2) Cell division in cortical cells of the root apex.

Organelles in the columnar cells of the proximal tubule in the mouse kidney are distributed in a polarized manner with mitochondria and peroxisomes in the basal region, nucleus and GA near the middle, and lysosomes and large vesicles at the apex. ER, as revealed by cyto-

Figures 3, 4. GERL cisternae stained for acid phosphatase in granulosa cells of 12 day corpus luteum. Sample taken 6 hr after i.m. injection of prostaglandin F2 alpha (See McClellan et al., 1977, for details). This treatment accelerates formation of dense bodies (DB), and autophagic vacuoles (AV) containing acid phosphatase, as part of the induced regression in the corpus luteum. In these 0.5 m sections, which have not been counter stained with heavy metals, lysosomes are prominent within the tubular network. V, small vesicles forming along tubular network. Fig. 3 X 12,000; Fig. 4 X 15,200. Both stereo pairs at 15° tilt.

chemistry for the marker enzyme glucose-6-phosphatase (Leskes et al., 1971), is distributed throughout the cell but follows the polarized distribution of other organelles (Fig. 5). In the basal region ER consists of interconnected laminar cisternae with numerous folds, following the contour of the invaginated PM and mitochondria, as well as extending toward the middle of the cell. As ER traverses through the cell toward the apical region, tubular branches extending from the laminar cisternae circulate around microvillar crypts (MC) and apical vesicles (V) (Fig. 5). These observations differ markedly from those of Favard et al. (1973) for ER in liver parenchyma where organelle polarity is not as prominent.

Recent studies on cell division in roots of young seedlings have shown that elements of ER and NE form a suprastructure around the spindle apparatus which excludes most other cytoplasmic organelles with the exception of segments of ER extending from the poles to the equatorial plane (Hepler, 1980, 1982; Favard, 1980; Hawes et al., 1981). Several earlier HVEM studies used samples stained by osmium impregnation, a treatment not selective for ER but which stains the forming or cis cisternae of GA, ER, NE and vacuolar compartments (Favard 1980; Hawes et al., 1980; Poux et al., 1974). The osmium-ferricyanide (OsFeCN) reaction, a method for selective staining of ER (Forbes et al., 1977; White et al., 1979), has been applied successfully to plant cells (Hepler, 1981), the exception being that uneven staining occurs in some cells (Schnepf et al., 1982).

The major changes in ER associated with mitosis

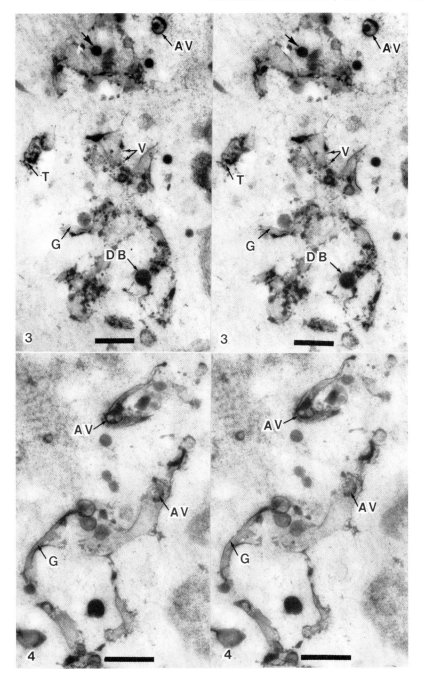

Figure 5. Section through proximal tubule of 2 week-old
Cf 1 mouse. Tissue stained for glucose-6- phosphatase.
Reaction product is present only in the NE and ER. In
this unstained 0.5 μm section details of ER organization
within several columnar cells of the proximal tubule are
readily observed. Along basal regions of each cell
laminar elements of ER follow the contour of the
invaginated PM and mitochrondria (M). Laminar ER
connects with tubular ER near the apex of each cell.
Tubular ER (arrow) extends into areas around microvillar
crypts (MC) and apical vesicles (AV). X 8,190; 30°
tilt.

Figures 6-11. Cell division in cortical cells of cotton
radicles from 18-hr-old seedlings. Stained with OsFeCN.
All sections, except Figs. 10 and 11, are 0.25 m and
counterstained with uranyl and lead salts.

Figure 6. As metaphase chromosomes align along plane of
division, numerous strands of tubular ER (arrows) form
an ellipsoidal lattice separating mitotic events from
the rest of the cytoplasm. Strands of ER extend from
poles toward aligned chromosomes (Ch) with some evidence
of interconnections of ER strands from opposite poles in
the space between adjacent pairs of chromosomes. LB,
lipid bodies. X 5,600; 40° tilt.

emerging from studies using OsFeCN staining represent a
good reference for comparison to cellular events involving
ER during the first divisions of cells in the radicle apex
of germinating seeds. Between 15 and 18 hr of germination
cell division begins in cortical cells of the radicle apex
in cotton seeds, as part of the process of establishing an
active meristematic region. Because of a close relationship between lipid bodies and tubular ER seen at
earlier stages of germination, it is of interest to follow
this relationship during cell division and also to
determine whether ER is involved directly with various
mitotic events. During late prophase numerous elements of
ER and NE fragments form a zone of exclusion around the
condensed and aligning chromosomes. Progression to
metaphase (Figs. 6, 7) is marked by elaboration of an
interconnected network or lattice of ER into an ellipsoidal basket with separate strands extending from the

Ultrastructural Histochemistry With HVEM / 253

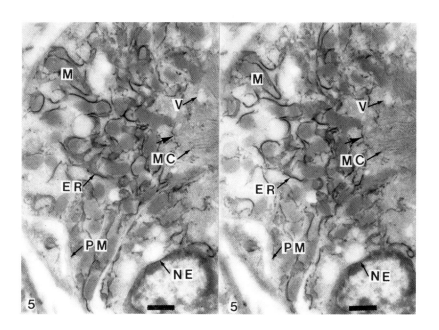

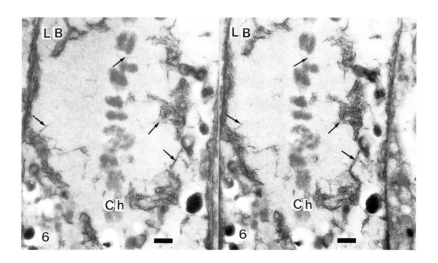

poles to the equatorial plane where some connect between metaphase chromosomes (Fig. 6).

Figure 7. Serial section of Fig. 6. The distribution of cellular organelles around the spindle complex is distinctive: Plastids (P) and mitochondria are present along the edge of the complex; Clusters of lipid bodies (LB) are present in cell corners; Protein bodies are situated between clusters of LB; Strands of ER (arrows) surround LB, form a spindle basket which separates chromosomes (Ch) from the rest of the cell, and run along cell wall (CW). x 6,000; 40° tilt.

These pole-derived strands of ER have two functions: one is to provide a calcium-rich membrane component (Wolniak et al. 1981) running parallel to the track of spindle microtubules along which chromosomes move during anaphase (Hepler et al., 1981; Hawes et al., 1981); the other is to form a lattice mesh network at the equatorial plane during telophase for entrapping Golgi vesicles which fuse and become part of the cell plate complex (CP) of cytokinesis (Fig. 8, Hepler, 1982). The CP expands uniformly and continuously by vesicle fusion between newly added strands of ER from NE and peripheral clusters of lipid bodies (Fig. 9). Advancing edges of CP are formed in an ordered fashion through the assembly of a double scaffolding of ER. One portion of the scaffolding is at the advancing edge of CP and has a web-like structure (Fig. 9); the other component is a loose lattice of ER running parallel to CP, and connecting to the latter by

short perpendicular strands of tubular ER at periodic intervals (Fig. 9). This intricate double scaffolding appears integral to formation of CP and cell plate expansion along the plane of partition for formation of two daughter cells (Figs. 9-11). This double ER scaffolding also appears to function as a stabilizing and regulating structure involved in development of the advancing CP through synchronous addition and fusion of Golgi vesicles.

Upon fusion with the cell plate, contents of Golgi vesicles react positively to OsFeCN treatment. This observation has been interpreted to result from calcium present within the cell plate following calcium-induced

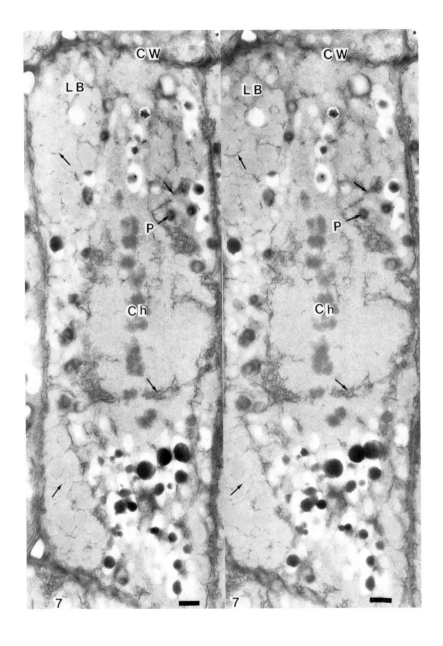

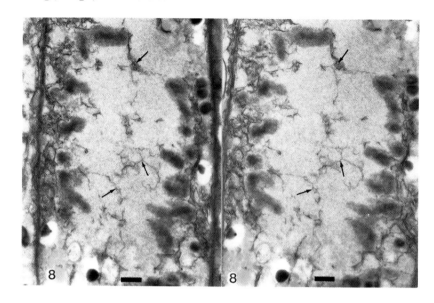

Figure 8. Late anaphase. Chromosomes at poles embedded in mass of ER. Some polar strands of ER extending toward the equatorial plane, connect with opposite strands (arrows) and form a lattice network. X 6,210; 30° tilt.

Figure 9. Telophase. Nucleus reformed in each daughter cell has a prominent nucleolus (Nu) and several chromosomes. Segments of ER extend from NE to cell plate and to ER along the cell wall (small arrows). Running parallel to the advancing cell plate complex (CP) is a secondary lattice of ER (large arrows) connected to the latter by perpendicular strands of ER. Formation of the secondary lattice appears to involve ER association with neighboring lipid bodies (LB). X 5,750; 20° tilt.

fusion of Golgi vesicles (Hepler, 1982; Wick and Hepler, 1980; Hepler et al., 1981). Diffuse staining of the cell plate, however, does not interfere significantly with visualizing the apposed ER lattice network (Figs. 9-11). ER remains associated with the cell plate and subsequent cell wall through formation of plasmodesmata, as entrapped vesicles fuse around ER cisternae traversing the lattice network.

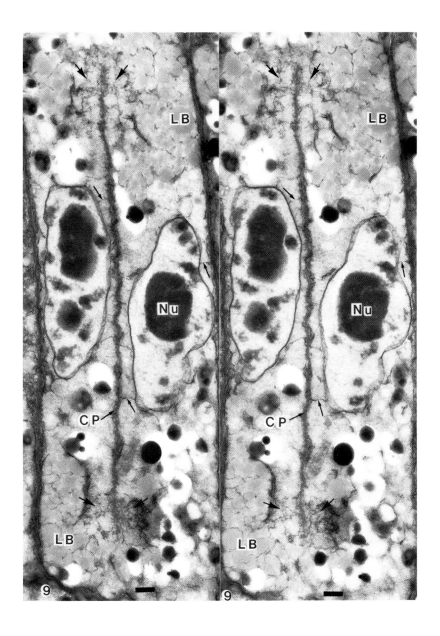

Figures 10, 11. These stereo pairs are from the same 2 m section but at different angles of tilt. The position of the nucleus (N) in each cell appears to be maintained by the large number of connecting strands of ER from the NE to ER along the CW, to clusters of lipid bodies (LB) (arrows), and cell plate (CP) (Fig. 11); LB occupy quandrant positions in the cell with ER extending to CP and NE (Figs. 10, 11). The CP has a permanent tubular lattice or webbing of ER formed at the advancing edges of CP (arrows, Fig. 11), as part of cytokinesis. Fig. 10 X 7,000; 8° tilt (-20°, -28° from horizontal). Fig. 11 X 9,000; 2° tilt ($+20^o$, $+22^o$ from horizontal).

CONCLUSIONS

The three-dimensional organization of membranous organelles visualized by combining selective staining methods with HVEM illustrates how effective and useful this approach is for acquiring new information on modulation of organelle structure in animal and plant cells. The new vistas made possible with enzyme cytochemistry and HVEM are unlimited and their exploration will clearly advance our knowledge on cellular dynamics. The acquisition of new information on organelle structure and function, examples of which are illustrated in this report, has proven important in refocusing our thinking on the existence of intricate and complex processes, manifested by dramatic temporal changes in different membrane compartments.

ACKNOWLEDGMENTS

The authors especially thank Christopher Pooley for printing and Eric Erbe for mounting and labeling the stereo pairs. We also thank Drs. Porter and Fotino, and Ris and Pawley for making the HVEM facilities available in Boulder, CO, and Madison, WI, respectively. These facilities are supported in part by grants from the Division of Biotechnology of NIH. Scientific Article No. A-3877, Contribution No. 6857 of the Maryland Agricultural Experimental Station.

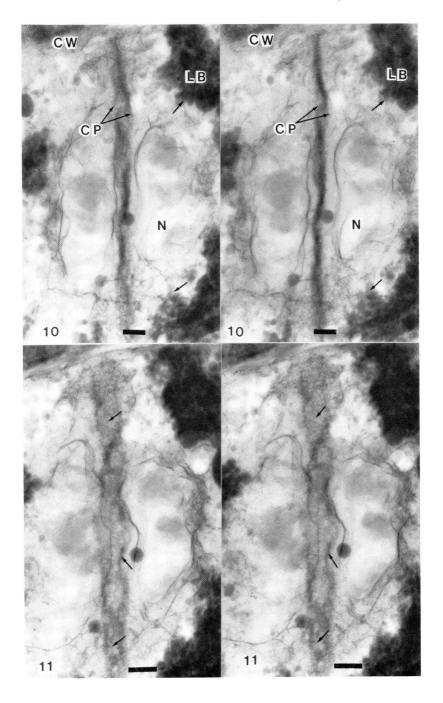

REFERENCES

Anderson WA (1972). The use of exogenous myoglobin as an ultrastructural tracer. J Histochem Cytochem 20:672.

Anderson WA, Vigil EL (1979). Myoglobin as an ultrastructural probe for studying permeability of the nephron. J Histochem Cytochem 27:1354.

Beeston BEP (1973). High voltage microscopy of biological specimens: some practical considerations. J Microsc (Oxf) 98:402.

Byers HR, Porter KR (1977). Transformation in the structure of the cytoplasmic ground substance in erythrophores during pigment aggregation and dispersion. I. A study using whole-cell preparations in stereo high voltage electron microscopy. J Cell Biol 75:541.

Carasso N, Delaunay M-C, Favard P, Pechaire J-P (1973). Obtention et coloration de coupes epaisses pour la microscopie electronique a haute tention. J Microscopie 16:257.

Cosslett VE (1971). High voltage electron microscopy and its application in biology. Phil Trans R Soc B Biol Sci 261:35.

Dupouy G (1968). Electron microscopy at very high voltages. Adv Opt Electron Microsc 2:167.

Favard P (1980). Recent aspects of development of high voltage electron microscopy. In: Electron Microscopy 1980 vol 4 High Voltage. Brederoo P, Van Landuyt J (eds) Leiden: Seventh European Congress on Electron Microscopy Foundation p 414.

Favard P, Ovtracht L, Carasso N (1973). Observations de tion of thick biological sections in the high voltage electron microscope. J. Microsc (Oxf) 97:59-81.

Favard P, Ovtracht L, Carasso N (1973). Observtions de specimens biologiques en microscopie electronique a haute tension. I. Coupes epaisses. J. Microscop 12:30.

Forbes MS, Plantholt BA, Sperelakis N (1977). Cytochemical staining procedures selective for sarcotubular system of muscle: modification and applications. J Ultrastruct Res 60:306.

Glauert AM (1979). Recent advances of high voltage electron microscopy in biology. J Microsc 177:93.

Goldfisher S (1982). The internal reticular apparatus of Camillo Golgi. J Histochem Cytochem 30:717.

Graham RC, Karnovsky MJ (1966). The early stages of absorption of injected horseradish peroxidase in the peroximal tubules of mouse kidney: ultrastructural cytochemistry by a new technique. J Histochem Cytochem 14:291.

Hama K (1976). Three-dimensional observations of the cellular fine structure by means of high voltage electron microscopy. In: Yamada E, Mizuhira V, Kurosumi K, Nagono T (eds): Recent Progress in Electron Microscopy of Cells and Tissues. Tokyo:University Park Press p 343.

Hand AR, Oliver C (1977). Cytochemical studies of GERL and its role in secretory granule formation in exocrine cells. Histochem J 9:375.

Hanker JS, Seaman AR, Weiss LP, Uena H, Bergman RA, Seligman, AM (1964). Osmiophilic reagents: new cytochemical principle for light and electron microscopy. Science 146:1039.

Harris N (1979). Endoplasmic reticulum in developing seeds of Vicia faba. A high voltage electron microscope study. Planta 146:63.

Hawes CR (1981). Applications of high voltage electron microscopy to botanical ultrastructure. Micron 12:227.

Hawes CR, Juniper BE, Horne JC (1981). Low and high voltage electron microscopy of mitosis and cytokinesis in maize roots. Planta 152:397.

Hepler PK (1980). Membranes in the mitotic apparatus of barley cells. J Cell Biol 86:409.

Hepler PK (1981). The structure of the endoplasmic reticulum revealed by osmium tetroxide-potassium ferricyanide staining. Eur J Cell Biol 26:102.

Hepler PK (1982). Endoplasmic reticulum in the formation of the cell plate and plasmodesmata. Protoplasma 111:121.

Hepler PK, Wick SM, Wolniak SM (1981). Structure and role of membranes in the mitotic apparatus. In: Schweiger, HG (ed): International Cell Biology 1980-1981 Berlin:Springer Verlag p 673.

Hermo L, Clermont Y, Rambourg A (1979). Endoplasmic reticulum-Golgi apparatus relationships in the rat spermatid. Anat Rec 193:243.

Hudson B (1973). The application of stereo-techniques to electron micrographs. J. Microsc (Oxf) 98:396.

Leskes A, Siekevitz P, Palade GE (1971). Differentiation of endoplasmic reticulum in hepatocytes. I Glucose-6-phosphatase distribution in situ. J Cell Biol 49:264.

Marty F (1973). Sites reactifs a l'iodure de zinc-tetroxyde d'osimium dans les cellules de la racine d'Euphorbia characias. C R Acad Sci Paris D 277:1317.

Marty F (1978). Cytochemical studies on GERL, provacuoles, and vacuoles in root meristematic cells of Euphorbia. Proc. Natl Acad Sci USA 75:852.

Marty F (1980). High voltage electron microscopy of membrane interactions in wheat. J Histochem Cytochem 29:1129.

McClellan MC, Abel JH, Niswender GD (1977). Function of lysosomes during luteal regression in normally cycling and PGF2alpha-treated ewes. Biol Reprod 17:499.

Novikoff AB (1964). GERL, its form and function in neurons of rat spinal ganglia. Biol Bull 127:358.

Novikoff AB (1976). The endoplasmic reticulum: A cytochemist's view (A review). Proc Natl Acad Aci USA 73:2781.

Novikoff AB, Novikoff PK (1977). Cytochemical contributions to differentiating GERL from the Golgi apparatus. Histochem J 9:525.

Paavola LG (1978a). The corpus luteum of the guinea pig. II Cytochemical studies of the Golgi complex, GERL, and lysosomes in luteal cells during maximal proges- terone secretion. J Cell Biol 79:45.

Paavola LG (1978b). The corpus luteum of the guinea pig. III Cytochemical studies on the Golgi complex and GERL during normal postpartum regression of luteal cells, emphasizing the origin of lysosomes and autophagic vacuoles. J Cell Biol 79:59.

Poux N (1981). Analyse de la cellule vegetale par microscopie electronique a haute voltage. Bull Soc Bot Fr 128:7.

Poux N, Favard P, Carasso N (1974). Etude en microscopie electronique haute tension de l'appareil vacouolaire dans les cellules meristematique de racines de concombre. J Microscopie 21:173.

Rambourg A, Clermont Y, Hermo L (1979). Three dimensional architecture of the Golgi apparatus in sertoli cells of the rat. Am J Anat 154:455.

Ris H (1969). Use of the high voltage electron microscope for the study of thick specimens. J. Microsc (Oxf) 8:761.

Schnepf E, Hausmann K, Herth W (1982). The osmium tetroxide potassium ferrocyanide (OSFECN) staining technique for electron microscopy. A critical evaluation using ciliates, algae, mosses, and higher plants. Histochem 76:216.

Simionescu N (1981). Transcytosis and traffic of membranes

in the endothelial cell. In: Schweiger HG (ed): International Cell Bilogy 1980-1981 Berlin:Springer Verlag p 657.

Simionescu N, Simionescu M, Palade GE (1973). Permeability of muscle capillaries to exogenous myoglobin. J Cell Biol 57:424.

Simionescu N, Simionescu M, Palade GE (1975). Permeability of muscle capillaries to small heme-peptides. J. Cell Biol. 64:586.

White DL, Mazurkiewicz JE, Barnett RJ (1979). A chemical mechanism for tissue staining by osmium tetroxide-ferrocyanide mixtures. J Histochem Cytochem 27:1084.

Wick SM, Hepler PK (1980). Localization of Ca^{++}-containing antimonate precipitates during mitosis. J Cell Biol 86:500.

Wolniak SM, Hepler PK, Jackson WT (1981). The distribution of calcium-rich membrane and kinetochore fibers at metaphase in lining endosperm cells of Haemanthus. Eur J Cell Biol 25:171

Wolosewick JJ, Porter KR (1976). Stereo high-voltage electron microscopy of whole cells of the human diploid line, WI 38. Am J Anat 147:303.

Wolosewick JJ, Porter KR (1979). Microtrabecular lattice of the cytoplasmic ground substance. Artefact or reality. J Cell Biol 82:114.

CATHODOLUMINESCENCE STUDIES: FROM CHROMATIN AND CHROMOSOME
TO HUMAN HAIR

Samarendra Basu
Research Scientist III
Scanning Electron Microscopy
New York State Police Crime Laboratory
State Campus, Bldg. 22
Albany, New York 12226

The biological stains, particularly the aromatic dye and drug compounds, have in common a much higher extinction coefficient in the ultraviolet and visible wavelengths of light than the macromolecules to which they bind. Therefore, if a dye (e.g. fluorescein, acridine orange, etc.), or a drug (e.g. lysergic acid diethylamide (LSD), quinacrine dihydrochloride, etc.), fluoresces under the ultraviolet light it should give effective cathodoluminescence (CL) in the scanning electron microscope (SEM). For biological studies these fluorochromes have to be applied to specimens at a suitable concentration (<5% W/V) of the dye (or drug) that will avoid concentration quenching. In fact, several studies performed in this manner have proved that the CL yield of stained cells and tissues is unexpectedly high. The emission is stable even at room temperature (25°C) and this has allowed recording of images of those specimens before 'fading' of CL due to electron bombardment became apparent (Pease, Hayes 1966; Barnett, Wise, Jones 1975; Barnett, Jones, Wise 1975). But these useful dye and drug compounds are still few (viz. thioflavine T, quinacrine dihydrochloride, acridine orange, ethidium bromide and fluorescein isothiocyanate) (for various references, see Basu 1980a, b.). The halogenation of aromatic structures may have an important role in electron resistance. This insight prompted an earlier CL investigation of the Giemsa stain and its application to chromosomes and chromatin (Basu 1980a, b). In this investigation (Basu 1980a, b) the CL images were obtained in the transmitted mode in an ETEC Autoscan SEM, using a system that combines a light pipe and a photomultiplier. The main results have been verified more recently in the surface emissive mode in an AMR 1000 SEM. The combined CL results are reviewed in this report.

   Because the fluorescent dye and drug compounds are generally electron sensitive, testing the stability of CL by a method that allows routine screening of possible CL compounds has been necessary. Several CL compounds including the potent, hallucinogenic drug, LSD, have been discovered as a result of this effort. It has been shown that maneuvering the CL contrast in the line-scanning mode could be a valuable aid in obtaining significant images. The main issue confronting the CL applications of biological stains is the need for clear identification of the source (i.e. dye or drug) of CL contrast. An unequivocal proof of this matter can be given by such studies as the dynamic (time-dependent) incorporation of a potential CL compound into a non-cathodoluminescent, highly impermeable biological object. Human head hairs and LSD have qualified for this experimentation in this report. Hairs of normal individuals were considered to be non-cathodoluminescent not only by a prior CL examination but also by their poor content of aromatic amino acid residues (total less than 6%), namely, tryptophan, tyrosine and phenylalanine (cf. West, Todd 1964).

MATERIALS AND METHODS

Stains and drugs, and their characteristics

   The commercially pure Giemsa Stain is a mixture of eosin-y, methylene blue and its azures A, B and C. The various sources of procurement of this stain mixture and other dye compounds are mentioned elsewhere (Basu 1980a, b). The tartrate salt of D-Lysergic acid diethylamide (LSD) or LSD-tartrate (US pharmacopeial) and cocaine hydrochloride (Merck & Co., Inc.) were pure drugs, used as standards for infrared and ultraviolet spectroscopies.

   Energy dispersive x-ray analysis of inorganic elements with an analyser (EDAX 707A) in the scanning electron microscope (AMR 1000) has indicated that none of these compounds except Giemsa contained the CL-quenching metallic (e.g. Fe) elements. The elemental compositions (%, W/W) were as follows: Giemsa-major ($>$10%) element Br; minor ($<$10%) elements S, Cl, and Fe ($<$1%); eosin-y or tetrabromo fluorescein - major element Br, no minor element; methylene blue- major elements S and Cl, no minor element; azure B- major element Br, minor elements S, Cl and Na; and fluorescein- major elements Cl, Na and traces of Ca (<1%). The light absorption and fluorescence characteristics of these compounds are given elsewhere (Holmes, French 1926; Conn, Holmes 1926; Stotz, Conn, Knapp, Emery 1950; Parker, Rees 1960; Basu 1980b). The absorption and emission characteristics of LSD and cocaine hydrochloride were known (Sperling 1972; Clarke 1974). The fluorescence maxima of these possible CL compounds are in the visible wavelengths of light (e.g. fluorescein at 515 nm; eosin-y at 540 nm; LSD at 438 nm (acid pH), 536 nm (alkaline pH, etc.).

Dye and drug screening for CL

The best substrate for CL screening is a polished, 13 mm diameter carbon planchet. The dye compound was applied as a droplet from a concentrated solution and then dried. Alternatively, the polished surface of the disk was soaked in a desired solvent (e.g. methanol) and this surface was touched to a spread, thin layer of granules of the dye. The CL emission was examined in the surface emissive mode using the CL detection system I (Fig. 1).

Chromatin and Chromosomes

The methods used to prepare chromatin and chromosomes on finder grids and their staining procedure with Giemsa (2.5% W/V) are presented elsewhere (Basu 1980b).

Sampling of hair, LSD incorporation and back-extraction

Because the naturally shed hairs are usually in the telogen (resting) phase, hairs in the anagen (growth) phase were plucked from heads of several individuals. The plucked hairs (20 of them), each of length 10-12 cm, were thoroughly washed by successive rinses in cycles, in three solvents, viz., distilled water, diethyl ether and acetone. Each rinse lasted for 5 min. and the time spent in each solvent was 20 min. The dried hairs, each encapsulated in a 20-lamda (λ) micropipette, were examined for their cleanliness and intactness in a stereo-microscope. Sample hairs containing a pigmented, plumpy root and attached sheath cells were selected.

These were adjusted inside the capillaries (20-λ) to pre-determined lengths to be soaked in aqueous LSD solution (0.3-0.6 mg/ml). The capillary rise to a distance of about 2.6 cm had to be considered to vary the soaking distance on individual hairs. The hair lengths extending out had to be immobilized on the capillary surface. The side of the capillaries containing the hair root was immersed in the LSD solution (1-2 ml) in a test tube. The test tube surface was precoated with a carbon paint to avoid bleaching (oxidation) of the drug by light. The mixture was allowed to stand at 60°C. The incorporation of LSD into hair is temperature and pH dependent. Otherwise, hair is virtually impermeable to drug molecules. The pH of the drug solution was adjusted with an acid (0.001N HCl, 0.001N HNO_3) or an alkali (2N NaOH) to a desired pH (pH 4.5-5.0 or pH 11.0-12.0). The optimal length of incubation (5 hr 60°C) was determined by withdrawing one capillary at a time and then stopping the drug incorporation by chilling (4°C), and then washing the hair in ice-cold distilled water, diethyl ether and acetone. This removes the drug deposits from the exterior of hair. After the final washes in acetone, the dehydrated hairs were examined for CL in an AMR 1000 SEM.

It must be added that most adhesive compounds including silver paint, carbon paint, rubber cement, and transfer tape, etc., are cathodoluminescent due to adhesive polymers present in them. These were avoided by mounting hairs on carbon planchets. Hairs were individually cut to lengths slightly longer than the soaking distance in LSD solution. Several such segments (2-3 cm) were mounted on a 2.5 cm diameter carbon planchet and fastened with a clipper. Longer hairs (>3 cm) were wrapped in rotations on carbon rods (diam. 0.6 cm). Each such rod was affixed at one edge to an aluminum pin (diam. 1.3 cm).

The hair segments representing maximum uptake (i.e. full CL views) were dissociated in strong alkali (5N NaOH, pH 14.0) in the presence of a protein denaturant (10 M Urea). Complete dissolution of hair required vortex mixing for 5 minutes. Usually LSD is extracted from street drugs with chloroform (Dihrberg, Newman 1966). The added mixture of chloroform and dissolved hair (1:1) was shaken for 5 min and then centrifuged at the highest available speed (3500 rpm) in an International Centrifuge. The denatured hair proteins (keratins) accumulate at the interface and the urea, unextracted by chloroform, remains in the upper aqueous phase. The chloroform layer was pipetted out from the bottom of the mixture and evaporated to dryness. The extract was re-dissolved in a smaller volume (100 microliter). A droplet of this was settled on a carbon planchet to examine the CL of the substance.

CL Examination

Two alternative systems of CL detection were employed (Fig. (s) 1 & 2). Before the CL detection system I (Fig. 1) was used, the SEM (AMR 1000) was properly aligned in the secondary electron mode, using the smallest possible aperture (diam. 100 µm) at the final condenser lens and a scintillator (ZnS) disk (AMRay, Bedford, Mass.) on the specimen stage at a tilt of 45° angle. With the specimen being focused (working distance 12 mm) the filament screws on top of the gun were gently turned to tilt the filament from its position of perfect alignment. This avoided registration of filament incandescence. After aligning the electron beam electronically, **the panel switch for electrical controls was turned off, and** the specimen chamber was opened. The Faraday cage collector and the scintillator disk of the photomultiplier (PMT) detector were removed. Also, the connectors (bias) for secondary and back scattered imaging were disconnected. Together with a secondary filter (yellow or green) in front, the nude light pipe of the PMT (EMI 6256B) was used as a direct photon collector (Barnett, Jones, Wise 1975). After pump down, the fluorescent grains of the scintillator were imaged at the smallest beam potential (5 KV) and then at the highest beam potential (30 KV). The corresponding line-scans were obtained, keeping the electron beam on the same area of the scintillator. Because scintillator is stable under electron bombardment, the line scans and those CL images were a guide to obtaining optimum results with the biological CL compounds in the same surface emissive mode.

The effective range of the photomultiplier (EMI 6256B with S13 photo-cathode) of this CL system (I) is approximately 360-600 nm with a maximum quantum efficiency (12.5%) around 420 nm (Parker, Rees 1960). The beam currents used with this system could not be given here as these were not measured.

The CL detection system II (Fig. 2) allowed transmission views of CL compounds and stained specimens. This system has

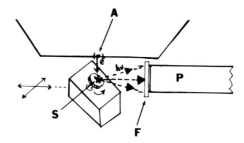

Fig. 1. CL detection system I: surface emissive mode. A: final condenser aperture (100 μm); F: color filter; P: photomultiplier; S: specimen; e: electron beam, and hy: CL emission.

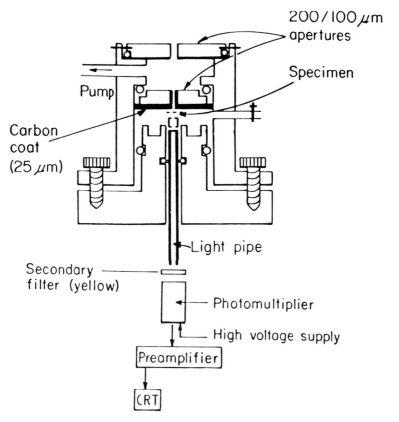

Fig. 2. CL detection system II: transmitted mode (from Basu 1980a)

been described in detail in a previous report (Basu 1980a). The main difference is that the top aperture assembly of the CL chamber was not shown in the earlier schematic representation (Fig. 1 in Basu 1980a). The apertured area of this assembly facing the specimen was coated with evaporated carbon (25 µm) to avoid back scattering of reflected electrons which may enter the quartz light pipe. The net effect of the apertures in the CL chamber and in the specimen mount below the grid was to reduce the input electrons per unit area of the light pipe. But depending upon impurities in them, the quartz pipes are excitable under ordinary conditions or after prolonged usage (Basu 1980a, b). Therefore, unless the dye emission is selected out with a color filter, lucite or perspex light pipes should be used instead of quartz pipes. All specimens in this report were examined uncoated as coating decreased observable cathodoluminescence.

RESULTS AND DISCUSSION

Dye and drug screening

Three examples of the dye/drug screening test are shown in Fig. (s) 3-5. These CL images of solid dye or drug were obtained in the surface emissive mode, using the CL detection system I (Fig. 1). The selection of electron beam potential (KV) depends upon the dye or drug. Both fluorescein (Fig. 3) and LSD (Fig. 4) can be excited to give CL emission at 5 to 10 KV or above whereas cocaine hydrochloride (Fig. 5) requires at least a 30 KV potential to emit light. Even at 30 KV the CL yield of cocaine is much inferior to that of LSD at 5 KV. The CL yield is directly related to beam current (not measured) that reaches the specimen. This beam current increases with the increase of electron beam potential. However, at higher electron beam potential (e.g. 30 KV) the CL contrast is lost faster due to 'fading' of CL caused by electron bombardment (cf. Fig. 3B vs Fig. 3C). The 'fading' of CL is more pronounced with LSD and cocaine than with fluorescein. With LSD and cocaine, complete quenching of CL in the irradiated area will occur in about 15 minutes (Fig. (s) 4B-D, Fig. 5B). Fluorescein is relatively more electron resistant and continues to emit light for about half-an-hour. In view of this electron sensitivity of dye or drug, the maneuvering conditions required for best imaging were obtained in advance by using the CL signal in the line-scanning mode (Fig. 3D). The marked spot sizes in

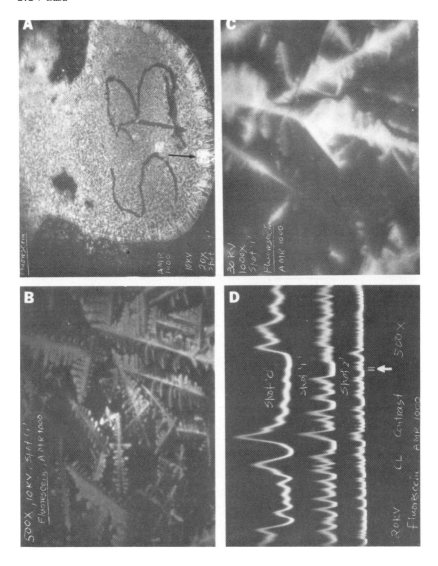

Fig. 3. CL images of dried fluorescein solution on a carbon planchet (A-C) and line scan of the CL signal (D) across the field in "B." CL records made in AMR 1000, using the detection system I in Fig. 1. Arrow in A: the area further examined in "B" and "C." Arrow in D: spatial separation of object elements in "B." Emission currents: 60-100 μA. C (small): contrast; S: distance.

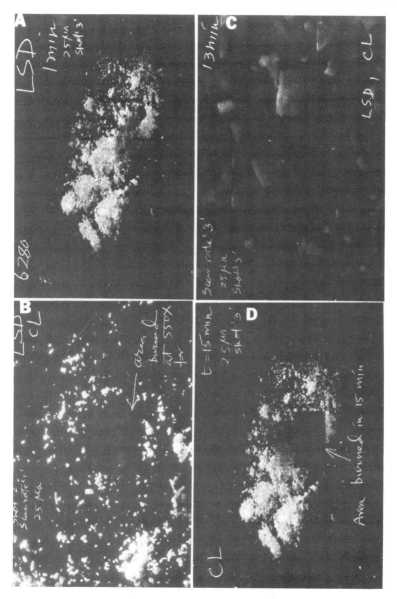

Fig. 4. CL images of solid LSD crystals on a carbon planchet at 5 kV in AMR 1000. A-D: Various magnifications of the same examined area. Mag in 'A' = 22X; Mag in 'B' - 110X, Mag in 'C' = 1100X and Mag in 'D' = 22X. Total irradiation time: 15 min. (See arrows in 'B' and 'D').

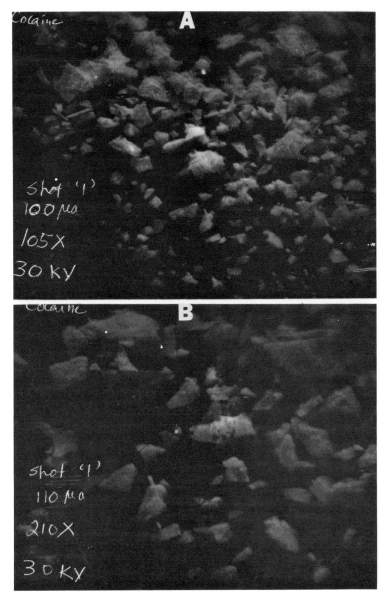

Fig. 5. CL images of cocaine-HCl crystals recorded at two magnifications in AMR 1000, using the CL detection system I in Fig. 1. Emission current (100-110μA) and other conditions are marked on the images.

this figure ('0', '1' and '2') refer to decreasing beam diameter in the SEM (AMR 1000) in this study. The CL contrast is represented by the height of the signal from the point on a base where it starts to rise. The spatial resolution at any spot size is revealed by the line separation of the CL peaks due to object elements on a line (unspecified) in Fig. 3B. Since beam current decreases as spot size is decreased, CL contrast is lessened as CL resolution is improved. A compromise between contrast and resolution is inevitable in the CL imaging technique. The extent of this adjustment is clearly indicated by a line-scanning before the sample area has the chance to burn out. Generally, one should try to use the lowest KV and the smallest spot diameter that will excite the CL emission. These two parameters together determine the excitation power input per unit area to the specimen. LSD is more luminescent than fluorescein and so a smaller spot size has been used with this compound (spot '2' or '3') than with fluorescein (spot '1'). This dye screening test has indicated that methylene blue (tetramethyl thionin) is non-cathodoluminescent whereas its alkaline oxidation product, azure B (trimethyl thionin), exhibits durable cathodoluminescence. The CL emission of rhodamine 6G is detectable but it is short lived. The CL yield of some dye compounds are weak when these are in solid form but their quantum yields may increase drastically as they bind to biological structures. Giemsa stain and its main ingredient, eosin-y, are in this category.

Chromosomes and chromatin

Fig. 6A and B represent the CL images of Giemsa-stained human A-group chromosomes prepared by acid-fixation. The chromosomes in Fig. 6A were spread on a Formvar (SiO-coated) supported finder grid and these were examined in the transmitted CL mode, using the CL detection system II (Fig. 2). In this particular case the top aperture (two apertures) assembly had to be taken off the goniometer stage to accelerate the search of chromosomes. This arrangement is shown in another report (see Fig. 1, Basu 1980a). The chromosome in Fig. 6B was spread on a polished carbon planchet and was detected by the CL detection system I in Fig. 1. The CL images of stained chromosomes in metaphase spreads taken with the CL system II have appeared elsewhere (Basu 1980a, b). The CL images in Fig. 6A & B clearly suggest that the CL contrast of bound Giemsa is appreciable (cf. CL image of Giemsa stain on carbon planchet (Fig. 6C). These acid-fixed chromosomes were entirely unsuit-

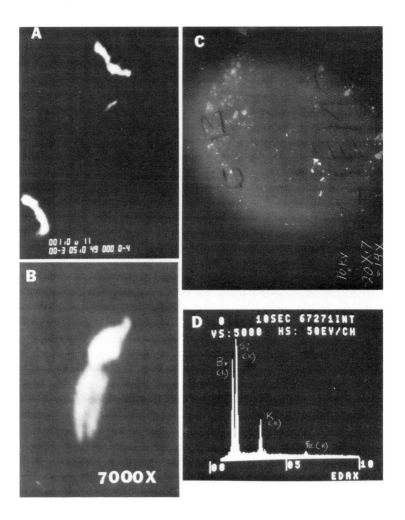

Fig. 6. Giemsa CL. A: transmitted CL images of two stained human chromosomes at 5 kV in ETEC Autoscan (from Basu 1980b); B: surface emitted CL image of one chromosome (A-group) at 5 kV in AMR 1000; C: Giemsa Stain (dried) on a carbon planchet (10 kV), and D: x-ray emission spectrum of the stained chromosomes in "A". Emission currents in B and C : 25-62 µA.

able for ultrastructural studies, as these chromosomes generally show no fine detail of their surfaces, not even in a transmission electron microscope (see Fig. 3 in Basu 1980b; Fig. 1a in Burkholder 1974). Qualitative analysis of inorganic elements with energy dispersive x-ray has shown that the major elements on the surfaces of Giemsa stained chromosomes are bromine (Br) and potassium (K), and the minor element (less than 1%) is iron (Fe) (Fig. 6D). No x-ray peaks due to phosphorus (DNA and RNA) and sulfur (chromosomal proteins) were detected. Among the detected elements in Fig. 6D, bromine (Br) is found in eosin-y and azure B, potassium (K) is presumably due to the buffer (Gurr buffer) and to the membraneous coat on the chromosome surface, and silicon (Si) is due to the background SiO coat of the Formvar substrate on the grid. Because the chlorine (Cl) and the sulfur (S) peaks of thionin dyes (methylene blue and its azures) are not observed, these must be present in negligible amounts on the stained, washed chromosomes. The CL emission of Giemsa is therefore mainly attributable to eosin-y and partially to azure B. It appears that the conventional banding treatments involving heat, enzymes, alkali, and high salt solutions may condition the chromosome surface to obtain Giemsa CL bands. These bands can be correlated with surface topography given by secondary electrons. This kind of correlative study would be impossible in the light microscope.

Further applications of Giemsa to isolated chromatin (Fig. 7) and nuclei (Fig. 8) had to be shown to prove the point that Giemsa is actually a nucleophilic (DNA & RNA-philic) stain. These results were obtained with the CL detection system II in Fig. 2. Thick fibers of granular (in-active) interphase chromatin of human meningioma (cultured cells) are shown in Fig. 7 (arrows 'Ch'). The width of these fibers is about 0.2 μm or less and these fibers also describe loop-like structures. The wet replication studies have revealed that chromatin fibers of this width consist of an irregular supercoiling of a 30-nm wide Finch and Klug (1976) fiber (arrow 'F', Fig. 9). This wet replica technique has been described in another report (Basu, this proceeding). These higher-order structures of condensed chromatin could have been overlooked without the prior evidence in the CL images of similar chromatin preparations (Fig. 7). This is the major advantage of CL microscopy over the conventional transmission electron microscopy in which fine structural analysis takes the precedence. The continuity of the chromatin fiber (arrows 'Ch') in Fig. 7 has been lost partially due to overcrowding of chromatin and

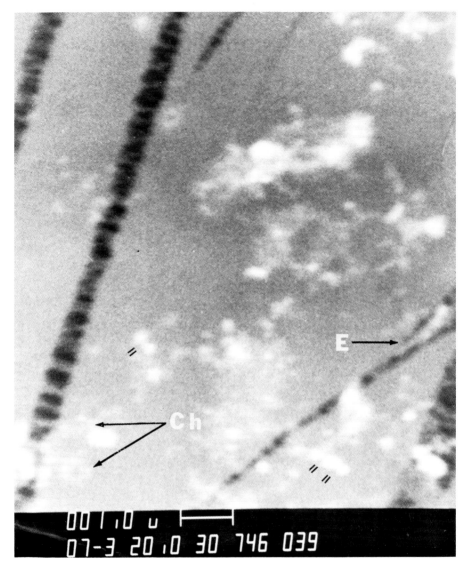

Fig. 7. CL image of surface-spread Giemsa stained human meningioma chromatin recorded in ETEC Autoscan at 20 kV. Emission current 175 μA beam current 0.4 x 10⁻¹⁰ A. Arrows: Ch, two successive loops of a 0.2 μm-wide chromatin fiber; E, end of a narrow crack on Formvar showing 60-70 nm separation; Parallel bars, 40-60 nm separation between chromatin granules (from Basu 1980b).

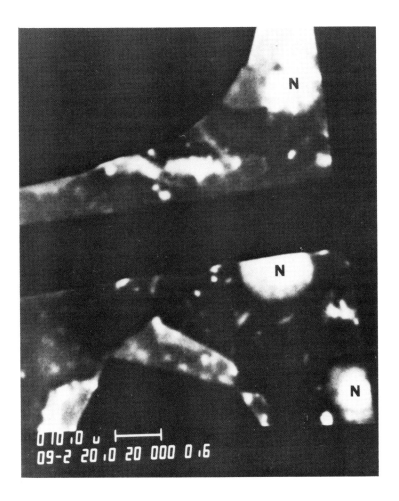

Fig. 8. CL images of three Giemsa stained nuclei (marked "N") on ruptured Formvar (SiO-coated) substrate on an electron microscopic grid. The torn areas of Formvar and the metallic grid bars are dark. Micrograph recorded in ETEC Autoscan at 20 kV; emission current 175 μA; beam current 2×10^{-10}A and electron dose 10^{-4} C/cm^2 (from Basu 1980b).

Fig. 9. Irregular superstructure of the Finch and Klug (arrow 'F') fiber (30 nm-wide) in wet SiO replicas of heterochromatin (cf. Basu 1979). Notice the thickness variation (arrows) like the vanishing width of the fiber 'Ch' in Fig. 7. CL did not resolve the 'F' fiber.

to uncoiling of these fibers into undetectible smaller chromatin fibrils as a result of stretching brought about by surface tension during drying (cf. Basu 1979).

For an important technical reason, the CL studies in the transmitted mode (CL detection system II) were performed with selected grids that exhibited a torn or a punctured substrate (Fig. (s) 7 and 8). The torn areas of Formvar did not give any indication of false cathodoluminescence in the light pipe underneath the specimen grids. The metallic grid bars were also dark corresponding in contrast with the torn areas of Formvar. This proved that there was no emission due to back scattered electrons. The CL emission was primarily from the stained nuclei ('N', Fig. 8), chromatin and chromosomes (Fig. (s) 6 and 7) and only in part from the excess unbound stain on substrate. The latter formed the background emission but did not affect the CL contrast of bound Giemsa. The Formvar films give a weak CL emission but that was also overcome by SiO coating.

In Fig. 7 luminescent granules of condensed chromatin were resolved up to a minimum distance of about 40 nm separating them (Fig. 7, bottom). The end of narrow cracks have been followed up to a width of 50 nm (arrow 'E', Fig. 7). This spatial CL resolution is expected to improve with field emission tips because it is ultimately the brightness of the electron source and the smallest beam diameter which together determine the excitability (i.e. excitation power per unit area) of CL and the resolution in CL images (see also Discussion in Basu 1980a). Where resolution is not of serious concern and detection takes the priority, the CL technology has no parallel to it. This capability of CL and the need for clear identification of the source (dye or drug) of CL contrast has prompted the studies to follow.

Incorporation of LSD into hairs

The results in this section refer to only in vitro tracer experiments which were conducted to measure the response of hairs to soaking in LSD solutions. The hair shaft is non-cathodoluminescent and is normally impermeable to LSD molecules at neutral pH. Incorporation of the drug occurs dynamically (time-dependently) at acid pH (4.5 to 5.0) and at alkaline pH (pH 11.0-12.0). Furthermore, the drug incorporation requires an activation energy which is provided by incubating the mixture (LSD-hair) at 60°C for 1/2 hr to 5 hrs. At the lowest incubation time (1/2 hr) the incorporation is partial. The CL images of these hairs (washed) indicate that the activated drug molecules first penetrate the cuticular scales covering the hair shaft (Fig. 10A, B). With longer incubation, the drug molecules move deep into the cortex and the medulla (cf. Fig. 10C). This behavior of incorporation is also observed at alkaline pH (Fig. 11A-C). The secondary electron images of these treated hairs indicate that the hair structure is rarely modified at the acid pH of the LSD solution. The alkaline solution of LSD partially digests the cuticular scales and as a result, the hair shaft appears to be smoother (Fig. 11C, D). The emitted signals were primarily from the soaked lengths of hair (see Fig. 11A). No emission due to reflected back scattered electrons from the unsoaked hair lengths were apparent. The incorporation of LSD into hairs was further confirmed by extracting back the drug from dissolved hairs. The droplet areas of the extract on carbon planchets showed microcrystals of LSD, similar to the epitaxially grown crystals of fluorescein in Fig. 3.

Several current lines of evidence indicate that the detection of LSD in human hair will soon be possible (Baumgartner, Jones, Baumgartner, Black 1979; Baumgartner, Jones, Black 1981; Viala, Deturmeny, Aubert, Estadieu, Durand, Cano, Delmont 1983; Ishiyama, Nagai, Toshida 1983).

Because only microquantities of LSD will be incorporated into hairs, depending upon the uptake, infrared spectrophotometry and mass spectrometry can be applied to the chloroform extract to confirm the presence of the drug. Mass spectrometry is sensitive enough to detect LSD quantities on the order of one microgram or even less. These studies and cross-sectional studies of hair are underway as a part of a research project in which the hairs of several subjects having definite history of drug abuse will be examined. Only hairs in the anagen phase of the hair growth cycle will be examined, in view of the potential interest of the correlation in time (i.e. distance from the hair tip) of the LSD intake with the CL features appearing in various single hairs. The CL signal in the line-scanning mode would be useful to this purpose of monitoring the drug abuse. The line-scanning technique has been successfully used for the detection of incorporated Cu-ions in human head hairs and to evaluate the spatial distribution of this tracer element along the length of hair (Basu, Alexander

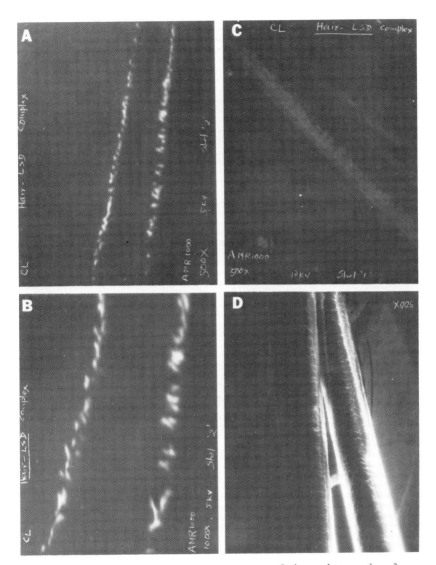

Fig. 10. CL images of LSD incorporated into human head hairs in acidic solution (pH 4.5). Incubation 1/2 hr. at 60°C. A-C: Segments of the same hair showing the progress of incorporation, starting with the cuticular scales (A) and then into the middle (i.e. medulla) of hair (C). D: Secondary electron images of the two hairs in "A" after the incubation period.

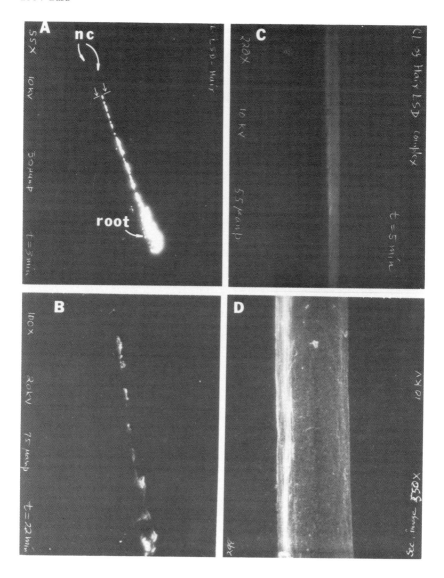

Fig. 11. CL images of LSD incorporated into human hair at alkaline pH (pH 11.5). A and B: Incubation 1/2 hr. at 60°C. C: Incubation 5 hr. at 60°C. D: Secondary electron image of a portion in "C". The smoothening of the hair shaft in alkaline solution was due to loss of the cuticular scales. Arrow "nc": non-cathodoluminescent, unincorporated portion of the hair in "A."

1983). The CL evidence of the drug can be given support if the CL spectrum of the drug is obtained directly from the photomultiplier output (Barnett, Jones, Wise 1975). The limitation of this approach will be revealed in multiple-drug abuse cases. The applications of thin-layer chromatography and gas chromatography could resolve this problem. Together these techniques can stand up in the courts of law and can render assistance to health and law enforcement agencies. The underlying reason is that drugs are often only detectable if a blood, or a urine sample, is collected within 24 hrs. of intake.

Although proofs of CL images of the hair-LSD complexes (Fig. (s) 10, 11) have been obtained for the first time with the CL-detection system I (Fig. 1), the collection efficiency of this system is still poor compared, for example, to the transmission type (Fig. 2), to reflection mirror type (Judge, Stubbs, Philip 1974), and to half-paraboloid, or ellipsoidal mirror type collectors, etc. (Hough, McKinney, Ledbetter, Pollack, Moos 1976). The CL detection system II is bulky and is suitable for studies in a SEM like the ETEC Autoscan. This collection system can hardly be used in the AMR 1000 scanning electron microscope. The collection system (I) (Fig. 1) needs to be improved or modified, so that the CL details can be resolved and the weak CL signals of other drugs (e.g. cocaine-HCL, heroin, etc.) can be processed. This task has to be considered along with the usage of low-dose sensitive films or emulsions and a beam chopping system.

Not only the forensic experts, but the experts in medical science will soon discover the impact of CL. After all, human head hairs are like the trunk of a tree: They store all information of personal hygiene including the abuse of potent drugs. The effects of these drugs extend all the way down to the chromosomes where they induce chromosomal damage (Cohen, Hirschorn, Frosch 1967; Egozcue, Irwin, Maruffo 1968, Nichols 1972). Whether the CL technology can be applied to early detection of these alterations would be a challenging task. The CL-banding of chromosomes must develop despite the many obstacles that currently stand in the way of the realization of this objective.

SUMMARY

Evidence has been presented to prove that cathodoluminescence (CL) studies of chromosomes and spread, Giemsa stained chromatin may lead to early detection of structural changes, such as the superstructure of heterochromatin. Several compounds including the potent hallucinogenic drugs, LSD and cocaine, have been discovered to be cathodoluminescent. The main issue confronting the CL applications using biological stains is the need for a clear identification of the source (dye or drug) of CL contrast. Unequivocal proof to this matter has been given by studies in which the time-dependent incorporation of LSD into the non-cathodoluminescent, virtually impermeable human hairs have been presented. The major implication of this finding is in forensic applications.

ACKNOWLEDGMENTS

The author thanks Martin Horan, Lt. J. Begley, Asst. Director and Mr. R. W. Horn, Director of New York State Police Crime Laboratories for their continued interest in the present work, and M. Larmour for preparing the typescript.

REFERENCES

Barnett WA, Jones EC, Wise MLH (1975). Observations on natural and induced cathodoluminescence from vaginal epithelial cells. Micron 6:93.

Barnett WA, Wise MLH, Jones EC (1975). Cathodoluminescence of biological molecules, macromolecules and cells. J Microsc 105:299.

Basu S (1979). Superstructures of wet inactive chromatin and the chromosome surface. J Supramolec Struc 10:377.

Basu S (1980a). Detection of cathodoluminescence of Giemsa stain and its applications. Rev Sci Instrum 51:435.

Basu S (1980b). Theoretical estimates of cathodoluminescence of Giemsa stained chromosomes. J Theor Biol 85:125.

Basu S, Alexander SJ (1983). The capability of the SEM-EDX to detect absorbed Cu-ions into human hair: A new application of line-scanning. Ninth Annual Meeting, Northeastern Association of Forensic Scientists, Oct. 7-8, 1983, Sheraton Heights, Hasbrouck, New Jersey.

Baumgartner AM, Jones PF, Baumgartner WA, Black CT (1979). Radioimmunoassay of hair for determining opiate-abuse histories. J Nuclear Med 20:749.

Baumgartner AM, Jones PF, Black CT (1981). Detection of phencyclidine in hair. J Forensic Sci 26:576.

Burkholder GD (1974). Electron microscopic visualization of chromosomes banded with trypsin. Nature 247:292.

Clarke EGC (1974). "Isolation and Identification of Drugs". London: The Pharmaceutical Press, p. 395, 789.

Cohen MM, Hirschorn J, Frosch W (1967). In vivo and in vitro chromosomal damage induced by LSD-25. New Engl J Med 277:1043.

Conn HJ, Holmes WC (1926). Fluorescein dyes as bacterial stains. Stain Tech 1:87.

Dihrberg A, Newman B (1966). Identification and estimation of lysergic acid diethylamide by thin layer chromatography and fluorometry. Analyt Chem 38:1959.

Egozcue J, Irwin S, Maruffo CA (1968). Chromosomal damage in LSD users. J Amer Med Assoc 204:122.

Finch JT, Klug A (1976). Solenoidal model for superstructure in chromatin. Proc Natl Acad Sci USA 73:1897.

Holmes WC, French RW (1926). The oxidation products of methylene blue. Stain Tech 1:17.

Hough PVC, McKinney WR, Ledbetter MC, Pollack RE, Moos HW (1976). Identification of biological molecules in situ at high resolution via the fluorescence excited by a scanning electron beam. Proc Natl Acad Sci USA 73:317.

Ishiyama I, Nagai T, Toshida S (1983). Detection of basic drugs (methamphetamine, antidepressants, and nicotine) from human hair. J Forensic Sci 28:380.

Judge FJ, Stubbs JM, Philip J (1974). A concave mirror, light pipe photon collecting system for cathodoluminescent studies on biological specimens in the JSM2 scanning electron microscope. J Physics E: Sci Instrum 7:173.

Nichols WW (1972). The relationship of chromosome aberrations to drugs. In Meyler L and Peck HM (eds.): "Drug-Induced Diseases", Vol. 4, Amsterdam: Excerpta Medica, pp. 60.

Parker CA, Rees WT (1960). Correction of fluorescence spectra and measurement of fluorescence quantum efficiency. The Analyst 85:587.

Pease RFW, Hayes TL (1966). Scanning electron microscopy of biological material. Nature 210:1049.

Sperling A (1972). Analysis of hallucinogenic drugs. J Chromatographic Sci 10:268.

Stotz E, Conn HJ, Knapp F, Emery AJ, Jr (1950). Spectrophotometric characteristics and assay of biological stains. Stain Tech 25:57.

Viala A, Deturmeny E, Aubert C, Estadieu M, Durand A, Cano JP, Delmont J (1983). Determination of chloroquine and monodesethylchloroquine in hair. J Forensic Sci 28:922.

West ES, Todd WR (1964). "Text Book of Biochemistry." New York: Macmillan, Chap 27, pp. 1161-1203.

ULTRASOFT X-RAY HISTORADIOGRAPHY AND NEW DEVELOPMENTS IN ULTRASOFT X-RAY IMAGING

Ronald R. Cowden
Department of Biophysics
Quillen-Dishner College of Medicine
East Tennessee State University
Johnson City, Tennessee 37614

A. HISTORICAL PERSPECTIVE

Ultrasoft x-ray historadiography enjoyed a brief span of interest in biomedical circles in the late 1940's, the 1950's and early 60's (see Beneke, 1966). While earlier investigators had experimented with the idea (see Goby, 1913), Engström (1946) and his co-workers at the Karolinska Institutet in Stockholm were the first to develop a systematic approach to this technology (see Lindström, 1955 for full theoretical treatment). Ultrasoft x-ray historadiography is based on the expectation that a "white"; i.e., polychromatic, x-ray source, usually using a copper target and 0.8-5.0 KV accelerating potential, will be mainly absorbed by the atoms, oxygen, carbon, nitrogen and hydrogen. Since these are the elements that compose most biological macromolecules, a registration of the ultra-soft x-ray absorption image under exposure conditions that produce a linear relationship between silver halide reduction (photographic darkening) and the sum of O, C, N, and H consititutes a measurement of "dry mass" of the structures in question (Fig.1). The image was generally registered in a vacuum of at least 10^{-4} torr. on a small piece of spectroscopic plate, usually through a pin-hole camera arrangement in which the hole was covered with a celloidin film onto which aluminum had been deposited by evaporation to blacken the film and thus prevent the glow of the electron gun filament from exposing the photographic emulsion.

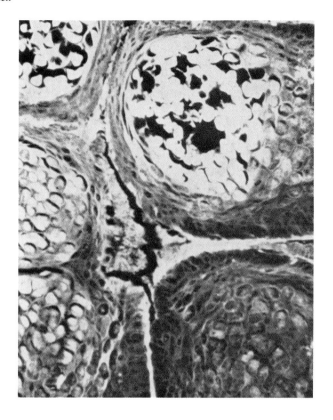

Figure 1: Microradiogram, X265, of a 5μm thick section of formalin fixed human epidermis with Molluscum contagiosum, registered at 1.5 kv.

The general design of these instruments is given in Figure 2. Note that the plate on which a dewaxed or cryostat section had been deposited was placed at some distance from the pin-hole, and that some form of manifold design was required which by-passed the pin-hole, thus avoiding rupture of the aluminum-coated, blackened celloidin film. The entire tube had to be evacuated to a vacuum of at least 10^{-4} torr. before the filament was turned on. Exposures with these systems were of the order of 20-40 minutes, and it was common to build a step wedge of celloidin, the elemental composition of which is

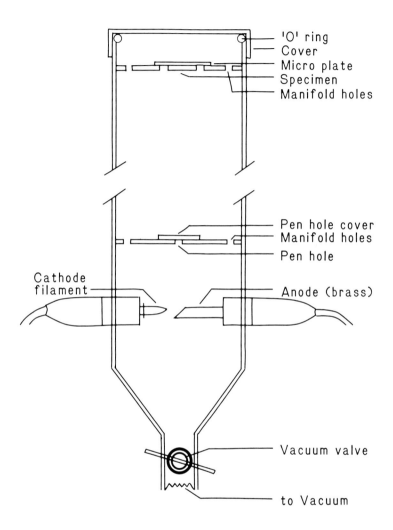

Figure 2: General design of an x-ray microradiographic tube and camera. The complete system requires a filament power supply and bias control, a 0-5 kv x-ray power supply and a fore-pump, oil diffusion pump and appropriate gages.

statistically similar to that of most biological macromolecules, for calibration. The whole procedure usually required at least two days of prior preparation: Gaevart-Lippman spectroscopic plate had to be cut into 2 X 2 cm. or smaller pieces, dipped in celloidin dissolved in amyl acetate and dried. Then a paraffin section of 4-6 mµ was floated onto the coated plate under a darkroom safelight, and again dried. The preparations had to be dewaxed unless cryostat sections were used, and dried again before final loading into the camera-x-ray tube for completion of an exposure. After the exposure of 20 to 40 minutes, the section was removed, the celloidin coating dissolved and the plate was developed with a high contrast developer such as Kodak D-19. The micro plate was then passed through "stop-bath", hypo and extensively washed. This plate could then be examined under the microscope with a high numerical aperture oil immersion objective. Under this set of conditions resolution 0.2 µm could be achieved. A historadiograph of Chironomus myofibrils is shown in Figure 3.

Figure 3: Microradiogram, X1500, of a 3µm thick section of a formalin fixed striated muscle of Chironomus, registered at 1.0 kv.

Quantitation of the plates was achieved by using a microdensitometer equipped with an x-y stepping stage, or by stepping the preparation directly in front of an x-ray sensitive photomultiplier tube mounted behind an aperture subtending 1.0–5.0μm (see Rosengren, 1959). This direct method did not produce the resolution that was possible with contact historadiographs registered on plates, but provided sufficient resolution for use in some specific neurobiology applications. By using x-ray sources at higher accelerating voltages, from about 5KV up to 20 KV, historadiographs were used for the demonstration of calcium deposition in conventional 5–10μm and "thick" (ca–100μm) ground sections of calcified tissues. If a sufficient flux of polychromatic x-rays could be generated, specific spectral regions could be selected using a bent crystal spectrometer. These x-rays were absorbed by the orbital electrons of specific higher atomic number elements. As an example, the nucleic acids were mainly demonstrated in sections of sea urchin oocytes when the distribution of phosphorus was imaged (Fig. 4 A, B, C and D).

This technology was tedious and time-consuming. In most systems, only nine to twelve historadiographs were exposed on a given working day, and much time was consumed breaking the vacuum, changing specimens, repumping the system and replacing torn aluminum-coated pin-hole covers. In some later designs, a number of plates could be loaded into a holder, the whole system evacuated, and either rotated or stepped into place after opening and closing a shutter after a specified interval of exposure. This development increased the convenience of exposure and the daily output of historadiographs. Ultrasoft X-Ray historadiographs were also prepared from smears of cells, squash preparations of dipteran polytene salivary chromosomes, sub-cellular structures (such as myofibrils, see Fig. 3) and cryostat sections.

In addition to the use of contact historadiography, the techniques for which were given in some detail by Beneke (1966), Cosslett and his group (see Engström, et al. 1957; Sharpe, 1979) developed a form of ultrasoft x-ray microscopy which made use of the magnification that could be obtained by the "distance squared" effect in which an image of a point source increases by the square of the distance from the source. This effect can be used to produce a magnified shadow of an object or specimen

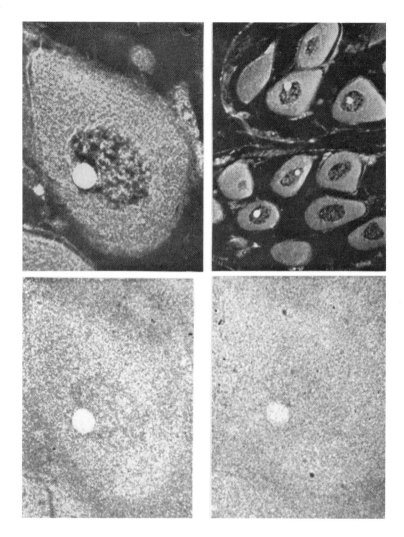

Figure 4: Microradiograms of sea urchin ovary fixed in ethanol-acetic acid (3:1), sectioned at about 8 μm A. Microradiogram at X195, registered at 1.5 kv continuous radiation. B. Same at X675, from the same section. C. Same section, X675, registered at 5.71 KXV images phosphorus (of nuclei acids). D. Same at X675 from the same section registered at 5.82 KXV no longer resolves phosphorus.

interposed at some point between the point source and the registration system. The useful magnification of these systems was limited and depended to a great measure on the elemental composition, thickness and symmetry of the specimen. Some attempts were also made, particularly by the Cambridge Group at Strangeways Laboratory which were summarized by Sharpe (1979) to develop lensing systems that would focus x-rays. These attempts met with some success and laid the foundations for some important later developments in X-ray lensing that followed the development of high-flux synchotron x-ray sources. In one design, magnifications of 150X were obtained. Cosselet's group was also instrumental in preparing the foundations of X-ray microanalysis.

The Dutch electronics firm, Phillips, entered this technology by offering an x-ray projection microscope version of their single lens electron microscope, the EM-75, and an ultrasoft x-ray historadiography unit with a Beryllium window which was designed to be used in air. Kodak (Rochester, N.Y.) developed a special un-perforated 35mm film for this system, and a punch was included with each unit that allowed the operator to punch out microfilms of the proper diameter. This apparatus designed with qualitative applications in mind, and was mainly used for quantitative work in the late Professor W. Sandritter's laboratory in Giessen (F.R.G.) in a modification in which a vacuum system was added (see Beneke, 1966). However, the anticipated interest in x-ray historadiography and x-ray microscopy did not develop, and these products were only offered for a few years in the late 1950's and early 1960's.

The development of practical versions transmitted light of interference microscopes in the early post-World War II period also contributed to the loss of interest in ultrasoft x-ray historadiography. Since interference microscopes could be used with living cells and the image immediately recorded by photography or photometry, it became the preferred technology for demonstrating the distribution of dry mass in biological systems within a relatively short period of time. However, the registration of "dry mass" by interference microscopy has not been without its problems. It is an inherently low numerical aperture, thus low resolution, method. Further, the precision of the method, which depends on the measurement of phase differences between the object and its "surround", requires accurate measurement of phase retardation or advancement. If this is done using a photometer, it is possible to achieve accuracies on the order of 100th of a wavelength, or if relative measurements are required, the system can be set to the linear portion of the sinusoidal (actually sine square) curve, and the difference between the object and its surround becomes a simple linear function of integrated absorbance (density) providing the refractive index of the "surround" is adjusted to within 0.25 λ of the optical retardation of the object (see Davies, 1958; Davies, Deeley 1956; Ross, 1967). In practice, measurements of dry mass in nuclei of Xenopus laevis endothelial cell nuclei have produced C.V.'s on the order of 50%, which was acceptable within the context of the experiments in question. However, without independent confirmation by another method, it is difficult to discriminate between measurement errors and true biological variation (Bereiter-Hahn et al., 1981, Bereiter-Hahn, 1984). While microscope interferometry does allow the estimation of dry mass distribution in living cells (see Bereiter-Hahn, 1984 this symposium), it does not ideally address the problem of registering dry mass distribution in sectioned material nor its measurement or demonstration at higher resolutions. Because the system depends on a determination of differences in refractive index and thickness, structures which exhibit birefringence can become major sources of errors and these effects must be compensated by orienting the preparation to avoid birefringence or by averaging errors (Galjaard, 1962). In any case, there are conditions under which interference microscopy does not offer the most satisfactory solution to the demonstration or estimation of "dry mass" in fixed, sectioned material; and in which some form of ultrasoft x-ray imaging might be preferable.

Interest in ultrasoft x-rays and their possible applications in biomedical sciences has revived since two new general classes of high flux sources have come into use in the past decade. The synchotron, used in a variety of high energy physics applications, generates a high flux of ultra-soft x-rays. Unfortunately for the investigator who might wish to use these in a biological imaging application, these systems are confined to major national **laboratories in the United States or similar installations** in Europe, and time on them—in most instances—has already been scheduled well in advance for the completion of the programs for which they were originally built. Most efforts at ultrasoft x-ray imaging with synchotrons have revolved around monochromation (or selection of wavelengths) and the development of complex lens systems to **focus x-rays with specific spectral characteristics. No doubt work along these lines will continue, but the average** university laboratory or research institute will simply not have reasonable access to these facilities.

In yet another development, it became obvious that plasma tubes could function as a source of high flux ultrasoft x-rays, and that images could be generated in a matter of seconds rather than minutes using these sources (see Epstein, 1984, Part B. of this section). While these systems were designed with mainly materials sciences applications in mind, it was inevitable that attention would, sooner or later, turn to possible biomedical applications. With the introduction of plasma sources as generators of ultrasoft x-rays, it has become possible to consider the configuration of systems that might be more generally available to the average university laboratory or research group. A system of this type is described by Epstein (1984) in Part B of this section.

Illustrations from Lindström (1955) reproduced by permission of Acta Radiologica, Stockholm, Sweden.

REFERENCES

Beneke, G (1966). Historadiography. In: Introduction to quantitative cytochemistry (G.L. Wied, ED). pp. 107-151. Academic Press, New York.

Bereiter-Hahn J, Wientzeck C, Brohl H (1981). Interferometric studies of endothelial cells in primary culture. Histochemistry 73:269.

Cosslett VE, Engstrom A, Pattee HA (1957). X-ray microscopy and microradiography. Academic Press, New York.

Davies HG (1958). The determination of mass and concentration by microscopy interferometry. In: General cytochemical methods (J.F. Danilelli, ED.) pp. 55-161. Academic Press, New York.

Davies HG, Deeley EM (1956). An integrator for measuring the "dry mass" of cells and isolated components. Exp. Cell Res. 11:169.

Engstrom, A (1946). Quantitative micro- and histochemical elementary analysis by roentgen absorption spectrography. Acta Radiol. Scand., Suppl. 63.

Galjaard H (1962). Histochemisch en interferometrisch onderzoek van hyalien kraakbeen. Thesis, Univ. of Leiden, Holland.

Goby, P (1913). Une application nouvelle des rayons x, La microradiographie. Compt. Rend. Acad. Sci. Paris, 156:686.

Lindstrom B (1955). Roentgen absorption spectrophotometry in quantitative cytochemistry. Acta Radio. Scand., Suppl. 125.

Rosengren BHO (1959). Determination of cell mass by direct x-ray absorption. Acta Radiol. Scand., Suppl. 178.

Ross KFA (1967). Phase Contrast and Interference Microscopy for Cell Biologists. St. Martin's Press, New York.

Sharpe RS (1979). Projection Microradiography. J. Microscop. 117:143.

ULTRASOFT X-RAY HISTORADIOGRAPHY AND NEW DEVELOPMENTS IN
ULTRASOFT X-RAY IMAGING

B. NEW APPROACHES TO ULTRASOFT X-RAY IMAGING

H.M. Epstein
Battelle Memorial Institute
Columbus, Ohio

Two types of soft x-ray sources, which are of considerable current interest, are direct outgrowths of thermonuclear-fusion research. While the goal of thermonuclear fusion is to obtain nuclear reactions in a magnetically or inertially confined plasma, with a minimum amount of x-ray emission, it is an obvious step to include efficient x-ray emitters in the hot plasma and produce an intense source of soft x-rays. The two initial fusion concepts that have been developed into laboratory x-ray systems are "Dense Plasma Focus" (DPF) and "Laser Fusion". The DPF consists of a collapsing magnetic field which compresses an ionized gas into a three-dimensional focal region. This basic concept has been modified by inserting puffs of a noble gas before each pulse to obtain intense bursts of K-line radiation from the gas. The laser-plasma x-ray source is related to laser-fusion only in that both rely on hot plasmas produced at the focus of a pulsed laser. In laser-fusion, a D-T filled sphere is uniformly irradiated with laser beams to produce compression of the central core. Whereas, most laser-plasma x-ray systems rely on the x-ray emission from the coronal plasma produced when a pulsed laser beam is focused on a large surface of higher atomic number material.

The relative merits of these two sources depend on the exact application. The laser-plasma sources of soft x-rays offer more flexibility in choice of spectrum, are higher in brightness, offer relative ease in bringing the x-rays out of the vacuum chamber, and produce less electrical noise.

On the other hand, the PUFF (see Shiloh et al., 1978) type DPF source is capable of considerably higher x-ray power per unit of cost, and produces that power in distinct K-lines. The discussions in this article will be limited to laser-plasma x-ray sources.

Because the temperature and density requirements of the plasma for a laser-plasma x-ray source are much lower than for a laser-fusion system, the required laser can be much smaller and less expensive than lasers for fusion. In fact, x-rays can be generated efficiently with mode locked laser pulses of several hundred mj (see Epstein et al., 1983). The main characteristics that differentiate a laser plasma x-ray source from conventional sources are:

(1) The x-ray spectrum comes from highly stripped species and predominantly L-line radiative or continuous in the kilovolt regime. Helium-like K-lines are also obtainable.

(2) The pulse width is very short, in the $\sim$ 0.1 to 10 nanosecond range.

(3) The source size is very small, $\sim$ 10 - 200 μm diameter.

This combination of characteristics makes possible several applications that have been unattractive with conventional laboratory x-ray sources. Most applications of soft x-rays fall into two categories: chemical analysis and x-ray microscopy. Chemical analysis includes: (see Epstein et al., 1984) Extended X-Ray Absorption Fine Structure (EXAFS) (see Mallozzi et al., 1981); spectroscopy of low atomic number elements; and Photon Stimulated Ion Desorption (PSID). While these chemical analysis techniques are certainly important for medical application, our attention will be confined to the microscopy applications.

The basic experimental configuration used in generating x-rays for near-contact microscopy is shown in the simplified sketch given in Figure 5. Typically, 1.06 micrometer wavelength neodymium laser pulses with energies ranging from .2 to 150 joules and pulse widths from .2 to 5 nanoseconds are focused onto a target to spot sizes ranging from 10 to 200 micrometers. Approximately 25 percent of the incident light can be converted into x-rays in the .3 to several keV regime. About 10 of the 25 percent lies between 1 and 2 keV.

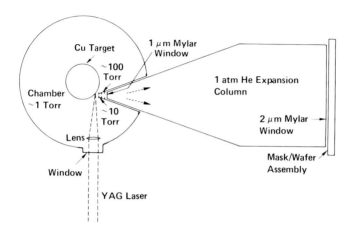

Figure 5: Basic Experimental Configuration

In this system, the target is a relatively large cylinder that advances on a helical drive to present a fresh, nearly flat surface area to successive laser pulses. Since each pulse destroys only about 10^{-4} cm^2 of target area, a cylinder with 100 cm^2 of surface area will produce about a million x-ray pulses. A liquid metal surface is also a viable alternative for a self-healing target surface.

The x-rays are extracted from the vacuum chamber and into a helium expansion column through a set of differentially pumped orifices, eliminating the troublesome problem of designing a vacuum window that transmits soft x-rays. The small source diameter permits an efficient extraction of a large cone of x-rays through the orifices. The first hole has a diameter of 1 mm and is located at a distance of 2 mm from the source.

The x-rays are emitted from a plasma layer with a temperature of $\sim$ 10,000,000 k (.86 keV). The x-ray spectrum consists primarily of line and recombination radiation to the K, L, or M shells depending on the choice of target material. Conversion efficiency of laser light to x-rays is highly dependent on the choice of target material as shown in Figure 6. To understand the reason for the peaks in this figure, it is helpful to realize that the L-lines are mostly caused by inelastic collisions between free electrons and ground state ions. The collisions excite bound electrons from the L-shell to the M-shell, and the x-rays are produced by the spontaneous radiative decay of the M-shell electrons back to the L-shell. Targets with Z above the copper peak have energy gaps between the L and the M-subshell that become

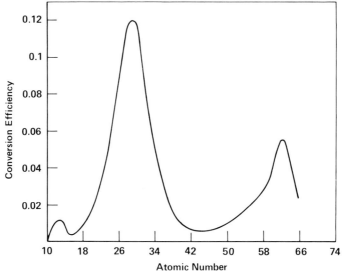

Figure 6. Dependence of X-Ray Conversion Efficiencies Above 1 keV on Atomic Number for 1.06 μm, 10^{14} w/cm^2

increasingly too wide to be efficiently excited. Targets with Z below the peak have energy gaps between the L and M subshell which increasingly fall below the 1 keV limit of the figure. These very soft x-rays may still be useful for high resolution microscopy of very thin specimens. Ultimately, a decrease of L-shell population below the peak caused a decrease inconversion efficiency regardless of the assumed threshold. The peak in K-shell conversion efficiency at a Z of about 13 can be explained in terms of

K and L shells in a similar manner. The K-shell peak is much lower in conversion efficiency, but the x-ray emissions are more energetic. An Al target emits predominantly He-like $K\alpha$ -line radiation at 1.6 keV. The K-shell efficiency can be increased substantially with shorter wavelength laser light, which penetrates to a denser portion of the plasma.

Since copper corresponds to the peak efficiency in Figure 6, it is the target usually preferred for microradiography. The spectral lines (Figure 7) are L-lines emitted from highly ionized species of copper. The band of L-lines can be moved up or down in energy, at some cost in efficiency, with targets of higher or lower atomic number.

The average x-ray power ultimately depends on the power of the laser. A 10 Hz 200 picosecond mode-locked YAG laser that produces 300 mj pulses is currently available.

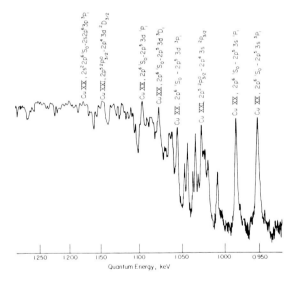

Figure 7. Densitometer Tracing of Bent Crystal Spectrograph of X-Rays Produced with Nd Laser Pulse

This 3 watt average power laser will produce $\sim$ .3 watts of x-rays over 1 keV. Slab amplifiers have been built, which will raise this output by 1 to 2 orders of magnitude.

RESOLUTION

The quality of a near contact microradiograph is determined by several factors: (1) geometric resolutions, (2) diffraction limitations, (3) mottle due to statistical fluctuations, (4) grain size and noise limitation in film or molecule size in photoresists, and (5) range of the emitted photoelectrons.

(1) The geometric resolution, δ_g, is determined by the finite source size, D, the source specimen spacing, R, and the effective distance between the specimen and the photoresist or film, L. Since $\delta_g = DL/R$ and the laser-plasma x-ray source is small (on the order of 10's of micrometers), the penubral blurring is small.

(2) Fresnel diffraction blurring, δ_F, is proportional to $\sqrt{\lambda L}$ when $R >> L$. The proportionality constant is on the order of 1.

(3) Mottle due to photon statistics is not normally a problem with the relatively insensitive photoresists, but it must be considered in the exposure of the x-ray films. Statistical fluctuations cause blurring (see Ter-Pogossian, 1967)

$$\delta_S \approx 2.5/C\sqrt{\pi n}$$

where C is the fractional contrast of the photons in the film plane, and M is the fluence in photons/cm^2 impinging on the film.

(4) The finite size of the grains in the film or the molecules in the photoresist form a limitation on the resolution because a whole grain or whole molecule is exposed.

(5) The range of the emitted electrons is only a limitation on resolution if detail smaller than $\sim$.1 micrometer is required. Then a trade-off between electron range and diffraction must be considered. Theoretical resolutions of less than 100 Å are possible with very soft x-rays (see Spiller et al., 1977).

APPLICATIONS

Several examples of radiographs and microradiographs are included to demonstrate the various capabilities of this type of x-ray source. The radiographs of the bee in flight (Figure 8a) and the beetle (Figure 8b) demonstrate the stop-action capability of the few billionth of a second x-ray pulse. The pictures of the meal worm (Figures 10 and 11) illustrate the use of a digitized scan to quantitatively analyze the salt concentration in small glands and ducts (Epstein et al., 1979).

Figure 8a. Bee in Flight

X-Ray Historadiography / 307

Figure 8b. Beetle

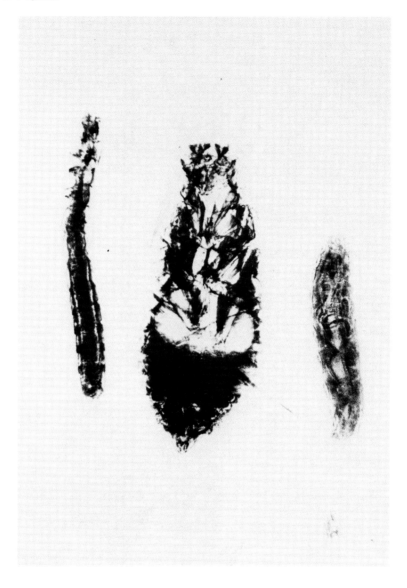

Figure 9. Meal Worm (10X)

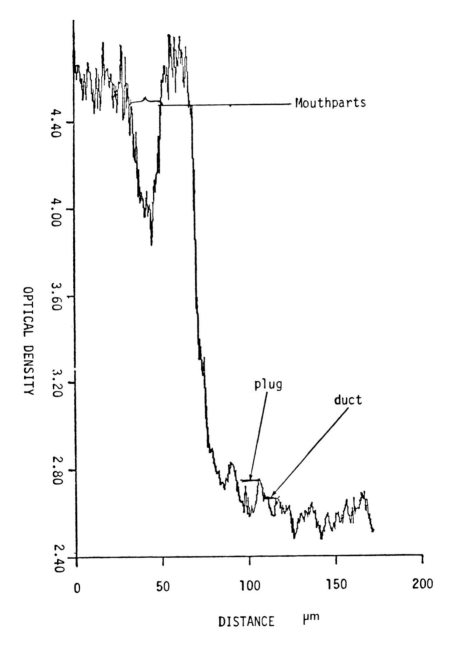

Figure 10. Digitized Scan of House Mite Radiograph

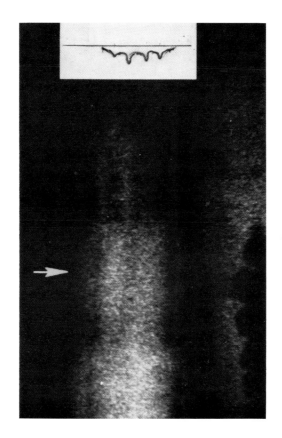

Figure 11. Digitized Scan of Radiograph of Rectal Complex of Meal Worm (20X)

REFERENCES

Shiloh J, Fisher A, Rostocker N. Phys. Rev. Lett., 40, 515 (1978).

Epstein H, Mallozzi P, Campbell B. SPIE, 385, 141 (1983).

Epstein H, Schwerzel R, and Campbell B. Laser Interactions and Related Plasma Phenomena, 6, edited by H. Hora and G. Miley, Plenum Press, New York, 149-164 (1984).

Mallozzi P, Schwerzel R, Epstein H, Campbell B. Phys. Rev. A., 23 (2), 23 1981).

Spiller, E, and Feder R. X-Ray Lithography, Chapter 3, edited by H. Queisser, Springer-Verlag (1977).

Epstein H, Duke K, and Wharton G. Trans. Amer Micros. Soc., 98 (3) 426-436 (1979).

Ter-Pogossian M. The Physical Aspects of Diagnostic Radiology Harper and Row, New York, 188-193 (1967).

SEGMENTATION OF WET ELONGATED MITOCHONDRIA INTO SPHERES

Samarendra Basu
Scanning Electron Microscopy
New York State Police Crime Laboratory
State Campus, Bldg. 22
Albany, New York 12226

From structural studies of mitochondria in the electron microscope it is obvious that these membrane-limited organelles are very sensitive to fixation, rinsing, storage, staining, dehydration and drying (Sjöstrand 1978). This mitochondrial sensitivity results from the greater lability of the inner mitochondrial membrane, the organization of which is highly complex inasmuch as it has to offer a favorable environment for the functioning of a variety of enzymes that participate in cell metabolism, including electron transport and oxidative phosphorylation (Chance, Williams 1956; Chance 1965). Conceivably the changes in the state of cell metabolism has been mainly attributed to the energy-linked swelling - contraction of the inner mitochondrial membrane (Hackenbrock 1966). No rationale has been given as to why the shape and size of mitochondria, as revealed in cell sections, vary so greatly from one cell type to another: spherical or nearly so in lymphocytes and liver cells and long filamentous or cylindrical in leucocytes and heart cells, etc. (Lehninger 1972; Gupta 1983).

Isolated mitochondria also exhibit marked changes in morphology and enzyme activities due to changes in osmolarity of the suspending medium (Douce, Mannella, Bonner 1973). Elongated, filamentous mitochondria have rarely been reported to have existed in the cell-free state; for example, after drying by critical-point (CPD) technique of Anderson (1951). This allows us to suspect that the micron-size spherical mitochondria of many cell types (e.g. potato, yeast, protozoa and liver cells, etc.) as obtained by this technique could be very unreal. As a preliminary attempt to this investigation, wet-unstained, unfixed mitochondria have been examined by an application of the wet replication technique (Basu, Parsons 1975; Basu 1982).

Wet replication is a direct transfer of the aperture-limited, differentially pumped environmental chamber techniques (Parsons 1974), to a vacuum evaporator. Because the replicas (SiO) of wet specimens can withstand the bombardment of electrons, the radiation damage problem does not arise. The physical aspects of this technique (shape replication, replication of water surfaces, etc.) have been described elsewhere (Basu, Parsons 1976; Basu, Hausner, Parsons 1976). The technique has recently revealed a supramolecular organization of nucleosomes in heterochromatin (Basu 1979 a, b). This report presents a comparison between replicas of wet and critical-point dried potato mitochondria.

MATERIALS & METHODS

Mitochondria were isolated from potato and purified at 4°C by the method of Douce et al (1973). The pellets of mitochondria were washed and suspended in a pre-cooled (4°C) medium containing 0.3 M Mannitol (or sucrose), 5 mM $MgCl_2$, 10 mM KCl, 10 mM phosphate buffer (pH 7.2). The mitochondria (approx. 150 mg. protein/ml) were well coupled and displayed State 3 oxidation rates of about 500 nanomoles of oxygen per min per mg protein for NADH. But isolated mitochondria become functionally crippled after 8-9 hr of storage (4°C) when oxidative phosphorylation comes to a halt due to depletion of ADP and endogenous substrate. Therefore while the mitochondrial suspension was standing at 4°C, wet replication and CPD experiments were performed alternately within 9 hr from the time of isolation. These experiments were begun after the first three hours of storage in order to account for swelling and dispersion of mitochondria. The wet replication procedure has been described in Fig. 1. Glow discharge facilitates controlled thinning of the aqueous layer on the wet specimens (Basu, 1979b; Basu 1982). The hydrophilic grid substrates, the buffer, portions of isotonic fixative (1% W/V glutaraldehyde, pH 7.2) and the ethanol-water mixtures (5%-100% V/V ethanol) required for dehydration in the CPD method, were all pre-cooled to 4°C. After the mitochondrial suspension was settled as droplets on several grids (4°C, 5 min), these were divided into two batches: one batch for wet replication and the other batch for critical-

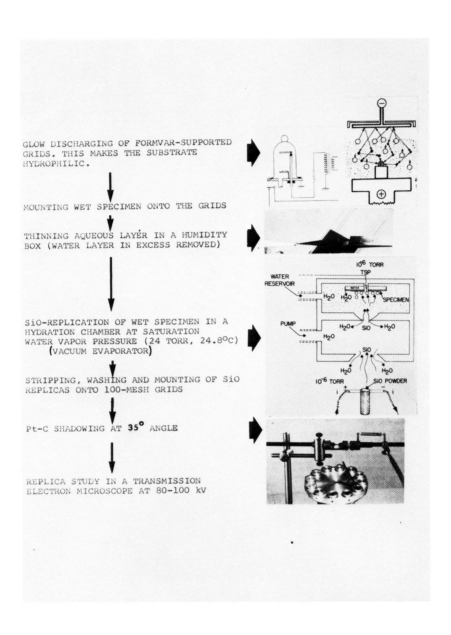

Fig. 1. The Wet Replication Procedure

point drying. The mitochondria for critical-point drying were used as unfixed, short-time fixed (5 min, 4°C) and long-time fixed (30 min, 4°C), following essentially Andrews and Hackenbrock (1975). After fixation, most of the specimen grids were processed through various gradients of ethanol and mixtures of Freon 113 and ethanol (10%-100% V/V Freon 113). Some of the specimen grids (unfixed) were also stained with 1% uranyl acetate during dehydration in alcohol gradients. Inside a CPD bomb the excess of Freon 113 was flushed away with liquified Freon 13 and the specimens were dried by taking them around the critical-point of Freon 13 (Cohen, Marlow, Garner 1968). These dried specimens were baked with evaporated carbon at 90° angle and then shadow-cast to obtain electron contrast, in the same manner as shown in Fig. 1 (bottom).

RESULTS AND DISCUSSION

Fig(s). 2 and 3 represent replicas of wet, unfixed mitochondria prepared at about 4 hr after isolation of mitochondria. The population consisted of elongated mitochondria of length up to 6 μm, mitochondria of intermediate lengths (2-3 μm) and many rounded mitochondria of diameter about 0.5 - 0.9 μm. These three different ranges of size were consistently found in replicas of wet mitochondria prepared up to 9 hr of storage at 4°C. There are clear indications in these micrographs that the larger mitochondria were actually precursors of the smaller forms. For example, the arrow heads in Fig. 2 indicate constricted areas along the length of two long mitochondria which have segmented into smaller forms (marked 1 and 2, Fig. 2). In Fig. 2 the mitochondrion marked d.mt is almost divided and is about to separate into two round forms. The arrow s.mt represents a typical round mitochondrion of diameter about 0.7 μm. This segmentation process is further suggested by the fact that the width of the elongated mitochondria usually matches the diameter of the round mitochondria in the same field (Fig. 3). These wet mitochondria represent lateral invaginations (arrows inv, Fig. 2) and occasionally cross-ridges of width about 50 nm (arrow cr, Fig. 2). These features are due to the folded, denser inner membrane (cristae, 30 nm-thick) because the covering outer membrane is known to be much thinner (7 nm-thick) (Sjöstrand 1978). Under similar conditions of wet replication, the surface polysaccharides of fungi **reveal themselves as 5-10 nm-wide fibrils (arrows, Fig. 4).**

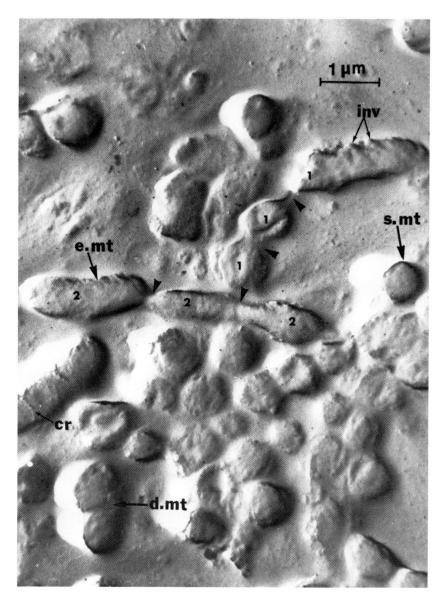

Fig. 2. Replicas of wet, unfixed potato mitochondria. The field shows segmentation behavior of elongated mitochondria (e.mt) (#1 & 2) of length 3-6 μm into the conventional round forms (s.mt & d.mt). See the text for other notations. Mag. 23, 450X.

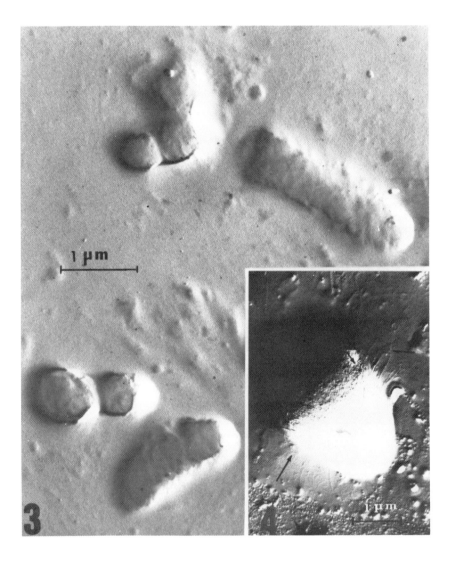

Fig. 3. Replicas of wet, unfixed potato mitochondria. Magnification: 31,500X.
Fig. 4. The replica of a wet, unfixed fugus (Trichoderma viride). Arrows show 5-10 nm polysaccharide fibrils protruding from the fungus surface. Magnification: 18,000X.

Obviously the surfaces of mitochondria do not reveal high resolution features on the order of intramembrane particles (width 10 nm).

The main observation is that wet potato mitochondria are found in diversified forms: elongated or filamentous (marked '2', Fig. 2), cylindrical-to-bag-like (Fig. 2 left, corresponding to arrow cr), slab-like (Fig. 3 top, right) and round-shaped, depending upon the state of segmentation. Perhaps conventional electron microscopy (e.g. thin sectioning) has observed portions of this picture. The replicas of critical-point dried mitochondria support this view (Fig(s). 5-7).

These preparations (Figs. 5-7) were made at about 3 hr after mitochondrial isolation, using the same stock suspension as used for wet replication (Figs. 2, 3). The fixed (5 min. 4°C), dried mitochondria (Figs. 5-6) were all spheres with diameter in the range of 0.4-0.8 μm. Spherical mitochondria were highly comparable in size with the round mitochondria in wet replicas (cf. Figs. 2,3). Because the search of these dried specimen grids at low magnifications (e.g. Fig. 5) did not reveal the segmenting mitochondria, it is conclusive that the wet, elongated mitochondria (Fig. 2) were not the products of fusion or aggregation of the smaller forms. On the contrary only these reduced, round forms can be obtained after CPD drying because the segmentation is nearly complete. Fixation of mitochondria does not prevent this segmentation process (cf. Fig. s. 5 or 6 vs. Fig. 7). In stained, dried mitochondria protrusions are discernible on the surfaces of these mitochondria (arrows pr, Fig. 7). Andrews and Hackenbrock (1975) interpreted them as bleb-like protrusions of the outer membrane. Unless these protrusions are artifacts of aggregation of non-specific, smaller vesicles, they raise a serious question about the shortcomings of the CPD method in membrane studies because these protrusions are not observed on the surfaces of wet mitochondria (Figs. 2,3).

In both size and shape these spherical potato mitochondria are not different from critical-point dried rat liver mitochondria (Andrews, Hackenbrock 1975) and protozoan mitochondria (Walker 1975), etc. The rationale of this observation is that mitochondria are bound to undergo segmentation or fission in unnatural environments. The inflated spherical shape and bleb-like membrane 'puffs' indicate melting of dehydrated lipids during the final transition

Fig. 5. Replicas of fixed (1% w/v glutaraldehyde, pH 7.2), critical-point dried potato mitochondria. The dried mitochondria were all spheres (diam. 0.4-0.8 μm). Magnification: 12,250X.

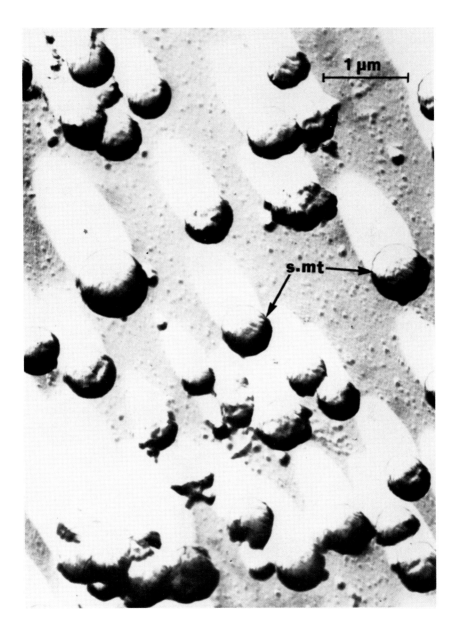

Fig. 6. Replicas of fixed, critical-point dried potato mitochondria at a larger magnification (35,000X). Arrows s.mt indicate spherical mitochondria.

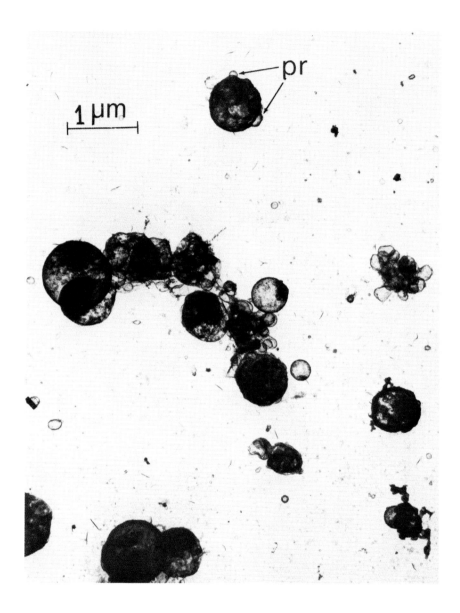

Fig. 7. Unfixed, stained (uranylacetate 1%), and critical-point dried potato mitochondria. Arrows pr indicate protrusions. Magnification: 31,500X.

(liquid-to-gas) of the CPD fluid. The warmer conditions necessary for embedding of sections have the same disadvantage. In fact it has been suspected that the warmer conditions of the wet replica chamber (chamber 25°C, mitochondria 4°C) could enforce the elongated mitochondria to segment. It is hoped that this interpretation will be verified with an improved wet replica chamber in which the temperature can be controlled to undertake wet replication at the temperature (4°C) of isolation and storage of mitochondria.

Several lines of support can be drawn to underline this lability aspect of mitochondria. Conventional alcohol and acetone dehydration introduce large changes in x-ray diffraction pattern of fixed membranes (e.g. myelin) by extracting lipids (Moretz, Akers, Parsons 1969). The transition temperatures of many lipids and lipoprotein membranes are well below the transition temperature of a standard CPD fluid (e.g. 28.9°C, Freon 13) (Ladbrooke, Chapman 1969; Hui, Parsons 1978). The inner membrane of mitochondria is more enriched in lipids than the outer membrane. The principal phospholipids of this inner membrane contain an enormous proportion (more than 50%) of unsaturated fatty acids. These fatty acids would drastically reduce the transition temperature of the inner membrane. The thin outer membrane may provide little resistance to segmenting forces derived from thermal melting and leaching of lipids from the inner membrane. Sjöstrand (1978) suggests that the mitochondrial inner membrane is continuous but consists of a two-phase system with an alternating aqueous phase and a non-aqueous mixed polar and non-polar phase. The latter is the matrix of two closely apposed inner membranes and their proteins. The surface of this matrix is impregnated at intervals by lipids which seal away the matrix from the surrounding aqueous phase. This model seems to be important because it is difficult to imagine that a segmentation force will chop the dense inner membrane into equal sized pieces unless such pieces were pre-existing in some sort of alternated, partitioned or even jointed structure inside the elongated mitochondria.

SUMMARY

Wet replication studies of potato mitochondria suggest that these mitochondria are basically elongated organelles of length 3-6 μm and width 0.5-0.9 μm. However due to unnatural extracellular and environmental conditions these

mitochondria are subject to spontaneous segmentation or
fission to yield organelles of smaller forms, and finally
to round mitochondria, the diameter of which equals the
width of the segmenting mitochondria. Only these reduced,
round mitochondria are found as spherical mitochondria by
the critical-point drying technique. Whereas this segmentation behavior of mitochondria points out the need for
further improvement of the wet replication technique, it
poses a serious issue: perhaps the real, undisputed shape
and size of mitochondria remain to be determined.

REFERENCES

Anderson TF (1951). Techniques for the preservation of three dimensional structure in preparing specimens for the electron microscope. Trans NY Acad Sci 13:130.

Andrews PM, Hackenbrock CR (1975). A scanning and stereographic ultrastructural analysis of the isolated inner mitochondrial membrane during change in metabolic activity. Expt Cell Res 90:127.

Basu S (1979a). Superstructures of wet inactive chromatin and the chromosome surface. J Supramol Struc 10:377.

Basu S (1979b). Evidence for superstructures of wet chromatin. In Nicolini CA (ed): "Chromatin Structure and Function," Part B, New York: Plenum Press, p 515.

Basu S (1982). New methods of investigating chromosome structure. In Barer R and Cosslett VE (eds): "Advances in Optical and Electron Microscopy", vol. 8, London: Academic Press, p. 51.

Basu S, Parsons DF (1975). The Wet replication technique. EMSA Bulletin 5:16.

Basu S, Parsons DF (1976). New Wet-replication technique. I. Replication of water droplets. J Appl Phys 47:741.

Basu S, Hausner G, Parsons DF (1976). New Wet-replication technique. II Replication of various wet specimens. J Appl Phys 47:752.

Chance B (1965). Reaction of oxygen with the respiratory chain in cells and tissues. J Gen Physiol 49:163.

Chance B, Williams GR (1956). The respiratory chain and oxidative phosphorylation. Advance Enzymol 17:65.

Cohen AL, Marlow DP, Garner GE (1968). A rapid critical-point method using fluorocarbons ("Freons") as intermediate and transition fluids. J Microscopie 7:331.

Douce R, Mannella CA, Bonner WD (Jr.) (1973). The external NADH dehydrogenases of intact plant mitochrondria.

Biochim Biophys Acta 292:105.

Gupta PD (1983). Ultrastructural study on semithin sections. Science Tools 30:6.

Hackenbrock CR (1966). Ultrastructural bases for metabolically linked mechanical activity in mitochondria. J Cell Biol 30:269.

Hui SW, Parsons DF (1978). Electron microscopy and electron diffraction studies on hydrated membranes. In Koehler JK (ed): "Advanced Techniques in Biological Electron Microscopy", Vol. II, Berlin: Springer-Verlag, p. 213.

Ladbrooke BD, Chapman D (1969). Thermal analysis of lipids, proteins and biological membranes. Chem Phys Lipids 3:304.

Lehninger AL (1972). Biochemistry, Chapters 17 and 18, New York: Worth Publishers, Inc., p. 365-393; p. 395-416.

Moretz RC, Akers CK, Parsons DF (1969). Use of small angle x-ray diffraction to investigate disordering of membranes during preparation for electron microscopy. Biochim Biophys Acta 193:12.

Parsons DF (1974). Structure of wet specimens in electron microscopy. Science (Wash) 186:407.

Walker GK (1975). The tubular cristae of protozoan mitochondria: Preservation by critical-point drying. Trans Amer Micros Soc 94:275

Sjöstrand FS (1978). The structure of mitochondria membranes: a new concept. J Ultrastruc Res 64:217.

ACKNOWLEDGMENTS

The present work was accomplished at Roswell Park Memorial Institute, Buffalo, New York and at New York State Department of Health, Albany, New York 12201. The author thanks Dr. Carmen Mannella for his comments, and M. Larmour for preparing the typescript.

REFLECTIONS AND CONCLUSIONS

RONALD R. COWDEN
DEPARTMENT OF BIOPHYSICS
QUILLEN-DISHNER COLLEGE OF MEDICINE
EAST TENNESSEE STATE UNIVERSITY
JOHNSON CITY, TENNESSEE

This symposium has attempted to highlight some of the advances in both light microscopy and biological ultrastructure. Probably the first observation one should make is that light microscopy is still alive and well, and progressing into new areas and applications. With the introduction of computers and sensors such as television cameras, scanned diode arrays and two dimensional diode arrays, we encounter an extensive capacity to manipulate microscopic images and rapidly convert digitized pixel data into objective formats. Similarly, as with the AVEC system which was discussed in this symposium, some substantial fraction of the inherent noise in lens systems can be removed from a conventional microscope image, leading to enhanced sensitivity, contrast and a capacity to quantify data from living cells. The con-focal microscope potentially achieves refinements in resolution that should be useful in any number of specific applications. Similarly, the large scale introduction of fluorescent dyes and reagents has made possible the development of a supravital quantitative cytochemistry which was exemplified in the presentation concerning membrane-bound calcium ions in cultured cells; there are a number of other options that can be directed toward categories of macromolecules, or specific organelles. It has also been demonstrated that both the advantages of the sensitivity and contrast of fluoresence microscopy can be utilized to study differences in nuclear organization that can be reported in objective, machine-sensible values. Finally, it has been demonstrated that the venerable Feulgen reaction for DNA can be manipulated to achieve enhanced accuracy and information concerning chromatin organization and texture. When this method is fastidiously employed in conjunction with a high-resolution instrument, adherance to the Beer-Lambert laws can be demonstrated over a range of four logarithmic orders.

While we may marvel at the emerging technologies of ultrastructure, we should take some comfort in the fact that light microscopy still provides the basis for some very important advances in microscope technology and has opened the way to new applications. There are potential biomedical applications in ultraviolet microscopy, infrared fluoresence microscopy, laser microbeam microsurgery and micromanipulation which have not been covered.

It must be obvious that the presentations in ultrastructure only scratched the surface of this complex and rapidly developing area of microscope technology. Perhaps we should apologize for leaving out accoustic microscopy, ion-etching, computerized reconstruction of fourier transforms to extract a clearer three-dimensional rendering of structure, and photoelectron microscopy just to cite a few new directions. However, there were limits on time and space, as well as problems with the identification of available and appropriate authors. In spite of these limitations, we have had examples of the application of histochemistry viewed in three-dimensional perspective which has been made possible by the development of high-voltage electron microscopy, of the use of x-ray microanalysis to analyse the ionic environment of the nucleus and its effects on chromatin and/or nucleosome organization, of the use of wet-replica techniques which allow preservation of near life-like geometry in biological systems, of the options presented by cathode-luminesence, and a sampling of the options that use of ultra-soft x-rays might offer.

Virtually all these advances in microscopy technology have in common some reliance on microprocessor or computer technology, coupled in most instances with detector systems that present two-dimensional digitizable images. If this technology continues along the pattern of recent history, we can expect more powerful and less expensive systems in the near future. However, we mainly hope that this symposium will introduce these technologies to a variety of biologists who tend to work with "unusual" material since with some conspicuous exceptions, the more advanced microscopy techniques mainly have found applications in the biomedical area on conventional mammalian or well-established invertebrate models.

We can be pleased with the new experimental options opened to us by these technologies, and reflect that the next decade or two should see even greater progress in these areas.

Index

Acheta domesticus and *Acrida conica*, Feulgen absorption cytophotometry, 154, 156
Acridine orange
  cathodoluminescence SEM, 265
  microphotometry, DNA and chromatin structure, 191–193, 197
Actomyosin-containing contractile ring, quantitative VIM, 73, 80
Adenylate charge, L cell response to malate, microspectrofluorometry, 51–57
Allen video-enhanced contrast. *See* AVEC entries
Allium cepa, Feulgen absorption cytophotometry, 147
Aluminum ions, microphotometry, DNA and chromatin structure, 194–195
AMAC IIIS electro-optical flow transducer (Coulter), 213–221
  argon ion laser, 215–216
  cf. cell volume analyzer, 215
  coefficient of variation, 214
  collection angle, 215
  fluorescence light scattering, 213
  morphological information, 213
  photon flux, 214
  propidium iodide, 215–216, 219–221
  Raji cells, 216, 220–221
Ambystoma mexicanum, Feulgen absorption cytophotometry, 141, 143, 146
Amiloride, 235, 236
7-Aminoactinomycin D, microphotometry, DNA and chromatin structure, 192–193
Ammoniacal silver reaction, microphotometry, nuclear structure and composition, 197
Amphioma means, Feulgen absorption cytophotometry, 140, 141, 143, 146, 150–151

AMR 1000 SEM, cathodoluminescence SEM, 265, 266, 268
Annular objective lens, scanning optical microscopy, 107
Anti-tubulin IgG, IIF, AVEC-DIC analysis, 14, 16, 17, 21
Apis mellifera, Feulgen absorption cytophotometry, 158
Ascaris, microphotometry, DNA and chromatin structure, 197
Ascites, malignant, microspectrofluorometry, 65
ATP, 82, 83
  trap (ethionine), 51, 52, 56, 58
Autofluorescence, simultaneous light and electron scanning microscopy, 130
Automated multiparameter analyzer for cells. *See* AMAC IIIS electro-optical flow transducer
AVEC-DIC microscopic analysis of cytoplasmic transport, 13–24
  bidirectional movement, 19, 20, 23
  Brownian motion, 13, 19, 22
  cell lysis, 16, 17, 20
  frame memory, 14–16, 24
  mottle, 14
  MTs, 13–14, 23
    MTLEs, 16, 18–21, 23
  protein composition, IIF with anti-tubulin IgG, 14, 16, 17, 21
  Rana pipiens keratocytes from corneal stroma, 14, 19, 23
  saltation, 13
  spherical cf. elongate particles, 17–19, 22
  transducers, 23
  ultrastructure, 22–23
  Zeiss Axiomat, 14, 23, 24
  see also Video *entries*
AVEC methods, 5, 8, 9, 10, 327
Azures A and B, microphotometry, DNA and chromatin structure, 190

331

Bee in flight, ultrasoft x-ray imaging, 305, 306
Beetle, ultrasoft x-ray imaging, 305, 307
Benzo(a)pyrene, microspectrofluorometry, 66
Birds, endangered, Feulgen absorption cytophotometry, genome size, 156–157
Bombyx mori, Feulgen absorption cytophotometry, 154, 157
Botany, simultaneous light and electron scanning microscopy, 129
Bromine, cathodoluminescence SEM, 277
Brownian motion, AVEC-DIC analysis, 19, 22
Buffalo rat liver (BRL) cells, inducible, microspectrofluorometry, 66
Bufo, Feulgen absorption cytophotometry, 139

Calcium, 254
  X-ray microanalysis, nuclear ionic environment, 225, 234–236
  see also Video intensification microscopy, quantitative, Ca^{++} role in cell division
Calmodulin, quantitative VIM, cell division, 73
Cancer
  diagnosis, 91, 92, 97–98
  DNA/nucleus, 92
  markers, simultaneous light and electron scanning microscopy, 125–127
Carcinogens, fluorescent, microspectrofluorometry, 66–67
Cathodoluminescence SEM studies, 265–286
  acridine orange, 265
  AMR 1000 SEM, 265, 266, 268
  chromatin and chromosomes, 267, 275–281
    bromine, K, and Fe, 277
    condensed chromatin, 281
    EM grid, 279, 281
    Giemsa-stained, 275–281
    heterochromatin, Finch and Klug fiber, 280
    meningioma, human, 277, 278
  cocaine hydrochloride, 266, 271, 274, 285

detection systems I and II, 269–271, 274, 285
dye and drug screening, 271–275
ETEC Autoscan SEM, 265, 276, 278, 279, 285
ethidium bromide, 265
FITC, 265, 271, 272, 275
Giemsa stain, 266–267, 275–281
  eosin Y, 277
heroin, 285
LSD, 265–268, 271, 273, 275, 281–285
LSD incorporation, hair, 266–269, 281–285
legal applications, 285
line-scanning mode, 285
photomultiplier tube, 265, 269, 270, 285
quinacrine dihydrochloride, 265
thioflavine T, 265
CCL 136 (rhabdomyosarcoma) cells, microspectrofluorometry, 57–59
Cell culture, quantitative VIM, 74
Cell division. *See* Video intensification microscopy, quantitative, Ca^{++} role in cell division
Cell lysis, AVEC-DIC analysis, 16, 17, 20
Cell motility. *See* Video *entries*
Cell plate, high-voltage EM, 254
Cellular junctions, rat diaphragm, endocytosis, high-voltage EM, 246–248
Cell volume
  analyzer, cf. AMAC IIIS electro-optical flow transducer, 215
  determination, microinterferometry, computer-assisted, 28, 39–42
  reflection interference microscopy, 40, 41, 43
  Xenopus epidermis cell, 40
Centhophilus stygius, Feulgen absorption cytophotometry, 154–156
Centrifugation, scanning microscopy, simultaneous light and electron, 122–124
Chemical analysis, ultrasoft x-ray imaging, 300
Chironomus striated muscle myofibrils, ultrasoft x-ray historadiography, 292
Chloride, X-ray microanalysis, nuclear ionic environment, 229, 234, 235

Chlortetracycline, 74, 78–81, 83
Chromatin
  condensation, Feulgen absorption cytophotometry, 152
  hetero-, 90–91, 280
  microphotometry, nuclear structure and composition, 187–190
  protein demonstration methods, 197–200, 202
  structure, X-ray microanalysis, nuclear ionic environment, 223–224
  condensation, 224
  see also under Microphotometry, nuclear structure and composition
Chromomycin A_3, microphotometry, DNA and chromatin structure, 192
Chromosomes
  interphase cf. mitotic, X-ray microanalysis, nuclear ionic environment, 223–224, 236
  metaphase, high-voltage EM, 252–253
Cocaine hydrochloride, cathodoluminescence SEM, 266, 271, 274, 285
Coenzymes, free vs. bound, microspectrofluorometry, 58–59
Colloidal iron (Immers modification), microphotometry, DNA and chromatin structure, 190
Compartmentation, NADPH, microspectrofluorometry, 48–51
Computer, Feulgen absorption cytophotometry, 174, 180, 182, 184
  data handling, microcomputer, 169
  see also Image analysis; Microinterferometry, computer-assisted, living cell
Conclusions and reflections, 327–329
Confocal LM, 105–110, 327–328
Copper target, ultrasoft x-ray imaging, 301–303
Corneal stroma, keratocytes, AVEC-DIC analysis, 14, 19, 23
Coulter. See AMAC IIIS electro-optical flow transducer (Coulter)
Cricket, Feulgen absorption cytophotometry, 154–155
Critical-point drying technique, 313, 314, 316, 319–323

Cryomicrodissection, 226, 227
Cyclic AMP, microspectrofluorometry, 65
CYDAC, Livermore Laboratories, 95
Cyprinus carpio, Feulgen absorption cytophotometry, 140, 145
Cytochemical staining, high-voltage EM, 244
Cytokinesis lattice, high-voltage EM, 254–258
Cytophotometry. *See also* Feulgen absorption microspectrophotometry (cytophotometry)
Cytoplasmic transport. *See* AVEC-DIC microscopic analysis of cytoplasmic transport

Daphnia, Feulgen absorption cytophotometry, 156
Dense plasma focus, ultrasoft x-ray imaging, 299, 300
  PUFF-type scans, 300
Densitometry
  IBAS system, microinterferometry, computer-assisted, 29, 30, 40
  ultrasoft x-ray imaging, 293, 303
Depth discrimination and depth of field, scanning optical microscopy, 107–110
Detectors, scanning optical microscopy, 110
Diaphragm, rat, endocytosis, high-voltage EM, 245–248
DIC. *See* AVEC-DIC microscopic analysis of cytoplasmic transport
Diffraction
  Feulgen absorption cytophotometry, 167, 169–174, 177, 179–181
  ultrasoft x-ray imaging, 304
Digitized scans, ultrasoft x-ray imaging, 309, 310
DMN, 51
DNA
  measurement, PBMN cells, simultaneous light and electron scanning microscopy, 125–127
  /nucleus, cancer diagnosis, 92
  reassociation kinetics, cf. Feulgen absorption cytophotometry, 157–158
  reference standards, Feulgen absorption cytophotometry, 138

see also Feulgen absorption microspectrophotometry (cytophotometry); Microphotometry, nuclear structure and composition
DNP, 51, 57
Drosophila, Feulgen absorption cytophotometry, 153
Drosophila melanogaster, 202
  Feulgen absorption cytophotometry pseudonurse cells, ovarian tumor mutants, polytene chromosomes, 139–140, 143, 152
  sperm, 140, 145, 151
Drug screening. *See under* Cathodoluminescence SEM studies
Dry mass
  microinterferometry, computer-assisted. *See under* Microinterferometry, computer-assisted, living cell
  ultrasoft x-ray historadiography, 289, 295–296

Eggs, sea urchin, quantitative VIM, 74, 82–83
Electron microscopy cf. OBIC method, 111–112
Electron probe. *See* X-ray microanalysis of nuclear ionic environment, electron probe
Electro-optical flow transducer. *See* AMAC IIIS electro-optical flow transducer (Coulter)
Endocytic vesicles, high-voltage EM, 246–248
Endocytosis, rat diaphragm, high-voltage EM, 245–248
Endoplasmic reticulum, quantitative VIM, 73, 80; *see also* High-voltage EM, ultrastructural histochemistry
Endothelial cells, *Xenopus* heart, computer-assisted microinterferometry, 39
Eosin Y, cathodoluminescence SEM, 277
Epidermis
  human, ultrasoft x-ray historadiography, 290
  Xenopus, computer-assisted microinterferometry, 40

Erythrocytes, chicken, Feulgen absorption cytophotometry, 170–171, 173, 177
  transform, 174, 176, 182, 183
ETEC Autoscan SEM, cathodoluminescence SEM, 265, 276, 278, 279, 285
Ethidium bromide
  cathodoluminescence SEM, 265
  microphotometry, DNA and chromatin structure, 195
Ethionine, 51, 52, 56, 58

Fast Black K salt, microphotometry, nuclear structure and composition, 198
Feulgen absorption microspectrophotometry (cytophotometry), 124, 127, 137–158, 167–184, 190, 195–196, 198, 328
  absorbance vs. translation into DNA amount, 138–139
    relative photometric values and percentage of genome size, 139
  Acheta domesticus, 154, 156
  Acrida conica, 154, 156
  Allium cepa, 147
  Ambystoma mexicanum, 141, 143, 146
  Amphioma means, 140, 141, 143, 146, 150–151
  Apis mellifera, 158
  average nuclear point absorbance and errors in nuclear integrated absorbance for image microdensitometry, 177–179
  Bombyx mori, 154, 157
  Bufo, 139
  Centhophilus stygius, 154–156
  computer, 174, 180, 182, 184
    microcomputer data handling, crickets and grasshoppers, 154–155
  Cyprinus carpio, 140, 145
  Daphnia, 156
  degree of chromatin condensation (compaction error), 152
  differences in local stain intensity, 168–169, 174–176, 181, 183
    chicken erythrocyte transform, 174, 176, 182, 183
  DNA reference standards, 138
  Drosophila melanogaster, 153, 157
    pseudonurse cells, ovarian tumor mutants, polytene

chromosomes,
    139–140, 143, 152
  sperm, 140, 145, 151
*Drosophila virilis, D. grimshawi,
    D. hydei*, 153
endangered bird species, 156–157
cf. flow cytometry, 144, 151, 184
Galleria melonella, 154
Gallus domestica, 140, 143–146, 148,
    154, 155, 157, 170–171, 173
genome size estimation, 137, 139,
    144–146, 153–155
Gossypium hirsutum, 147
Hadenoecus subterraneus, 154–156
human, 145
Hymenoptera wasps, 154
insufficient stain darkness, 176–177
  chicken erythrocyte:thymocyte
      integrated absorbance ratio,
      177
integrated absorbance measurement,
    179–184
  ^{3}H-thymidine incorporation correlation,
      182–183
  human WBC, 180–181
  mouse bone marrow cells, 183
  off-peak wavelength, 179–180
Lilium, 147
Locusta migratoria, 156
methods, 140–143
  cf. DNA reassociation kinetics,
      157–158
  instrumentation, 141–142, 158
  integrated absorbances, 142, 146,
      148, 149, 177
  variables, 138–139
  Vickers photomultiplier tube, 142,
      147, 169, 177, 184
mouse thymocytes, 170–173, 175, 176
Notophthalmus viridescens, 141, 146,
    149–150
optical errors, glare and diffraction, 167,
    169–174, 177, 179–181
  coefficients of variation, 172, 176,
      180, 182, 183
  optical density, 169, 179
Pediastrum boryanum, 147
Plethedon glutinosus, 141, 146, 149

Poecilia formosa, 140, 145
Pseudemus scripta, 140
Rana esculenta, 141
Salmo gairdineri iridens, 140–141,
    143–146, 148, 154, 155
simultaneous scanning LM and EM, 124,
    127
specificity and validity, 137
Tradescantia paludosa, 147
variability of non-histone proteins and
    nuclear volumes, 152
various plant species, 146–147
Vicia faba, 147
Xenopus laevis, 138, 139, 141–148, 154,
    155–158
Zea mays, 147
see also Microphotometry, nuclear
    structure and composition
Filtering mode, video-enhanced microscopy,
    10
FITC, cathodoluminescence SEM, 265, 271,
    272, 275
Flavins, microspectrofluorometry, 45, 46, 59
Flow cytometry, 211–212
  cf. Feulgen absorption cytophotometry,
      144, 151, 184
  cf. static, 211
  see also AMACIIIS electro-optical flow
      transducer (Coulter)
Flow diagram, microinterferometry,
    computer-assisted, 35–36
Fluorescein, microspectrofluorometry, 59
Fluorescence light scattering, AMACIIIS
    electro-optical flow transducer, 213
Fluorescent photosensitizers,
    microspectrofluorometry, 67
Fluorescent probes. *See* Cathodoluminescence SEM studies; specific probes
Fontainebleau School (France), 90
Frame memory, 7–8
  AVEC-DIC analysis, 14–16, 24
Frittilaria meleagris, 130
Furocoumarins, microspectrofluorometry, 67

Galleria melonella, Feulgen absorption
    cytophotometry, 154
Gallocyanin-chromalum, microphotometry,
    DNA and chromatin structure, 190,
    191, 196

Gallus domestica, Feulgen absorption cytophotometry, 140, 143–146, 148, 154, 155, 157, 170–171, 173
Genome size estimation, Feulgen absorption cytophotometry, 137, 139, 144–146, 153–155
Geology, scanning microscopy, simultaneous light and electron, 129
Germinal vesicle breakdown, 234–236
Giemsa-stain, cathodoluminescence SEM, 266–267, 275–281
Glare, Feulgen absorption cytophotometry, 167, 169–174, 177, 179–181
Glow discharge, mitochondria, 314, 315
Golgi apparatus, high-voltage EM, 244, 245, 249, 250
Gossypium hirsutum, Feulgen absorption cytophotometry, 147
G6Pase, 250, 252–253
G6P, microspectrofluorometry, 47, 48, 59
Grasshopper, Feulgen absorption cytophotometry, 154–155
Guard reaction, microphotometry, nuclear structure and composition, 197

Hadenoecus subterraneus, Feulgen absorption cytophotometry, 154–156
Haemanthus, quantitative VIM, cell division, 81, 82
Hair, human, LSD incorporation, cathodoluminescence SEM, 266–269, 281–285
Hall mass fraction method, 226
Heart endothelial cell, *Xenopus*, microinterferometry, computer-assisted, 39
HeLa cells, quantitative VIM, 78–83
Heroin, cathodoluminescence SEM, 285
Heterochromatin
  Finch and Klug fiber, cathodoluminescence SEM, 280
  image analysis, 90–91
High-voltage EM, ultrastructural histochemistry, 243–259, 328
  cytochemical staining, 244
  endocytosis, rat diaphragm, 245–248
    cellular junctions and endocytic vesicles, 246–248
  ER, 244, 245, 249–250, 252–258

calcium, 254
cell plate, 254
cytokinesis lattice, 254–258
G6Pase, 250, 252–253
metaphase chromosomes, 252–253
mitosis, 250, 252–256
mouse kidney proximal tubule, 249, 251–252
root apex, 249, 250, 252–253
spindle complex, 254, 255
Golgi apparatus, 244, 245, 249, 250
lysosome formation, GERL cisternae, 245, 247–251
  acid phosphatase, 247, 249
  corpus luteum granulosa cells, 247
  dense bodies, 249–251
mitochondria, 249, 250, 252–255
nuclear envelope, 244, 245, 250, 252–253
nucleolus, 256
peroxisomes, 249
plasma membrane, 245, 246, 250, 252–253
plastids, 254, 255
cf. serial reconstructions, three-dimensional, 243–244
specimen preparation, 245–246
Histones, microphotometry, nuclear structure and composition, 197–200
Historadiography. *See* X-ray historadiography, ultrasoft
Hoechst compounds, microphotometry, DNA and chromatin structure, 192
House mite, ultrasoft x-ray imaging, 309
Hydrogen, X-ray microanalysis, nuclear ionic environment, 234
Hymenoptera wasps, Feulgen absorption cytophotometry, 154

IBAS system, densitometry, computer-assisted microinterferometry, 29, 30, 40
IgG, anti-tubulin, IIF, AVEC-DIC analysis, 14, 16, 17, 21
Illumination, video-enhanced microscopy, 7
Image analysis, 89–98
  computerized video, 89–91
  computers and software, 96–97
  costs, 98

cytodiagnosis, cancer, 91, 92, 97–98
  DNA/nucleus, 92
  detectors and scanners, 93–96
    photomultiplier tubes, 95
    two-dimensional diode arrays, 94–95
    video cameras, 93–94, 96
  flow cytometry, 212
  Fontainebleau School, density and geometric data, 90
  heterochromatin, 90–91
  intensified video, 89–90
  leukocyte differential counts, 91–92
  microinterferometry, computer-assisted, 29, 34–35, 42
  microphotometry, nuclear structure and composition, 188
  microspectrophotometry (Zeiss), 91
  morphometry and neuroanatomy, 89
  preparation and stain, 97–98
  quantitative VIM system, 75–76
  stereology, 90
  video-enhanced microscopy, 3
    quality, 6
  see also Computers
Immunocytochemistry, scanning microscopy, simultaneous light and electron, 129
Immunofluorescence, indirect, anti-tubulin IgG, AVEC-DIC analysis, 14, 16, 17, 21
Intercellular flow, modulation, microspectrofluorometry, 59–65
Interference microscopy, 295; see also AVEC-DIC microscopic analysis of cytoplasmic transport; Microinterferometry, computer-assisted, living cell
Interphase cf. mitotic chromosomes, X-ray microanalysis, nuclear ionic environment, 223–224, 236
Ionic environment. See X-ray microanalysis of nuclear ionic environment, electron probe
Iron
  cathodoluminescence SEM, 277
  colloidal (Immers modification), microphotometry, DNA and chromatin structure, 190
Isocitrate, 55

Kidney, proximal tubule, mouse, high-voltage EM, 251–252

Laser
  argon ion, AMACIIIS electro-optical flow transducer, 215–216
  and scanning optical microscopy, 104
  ultrasoft x-ray imaging
    fusion, 299, 300
    Nd, 300, 301, 303
Leitz
  Mach-Zehnder type interference microscope, 29, 33–34, 37
  MPV2 system with CELANA software, 189
Lens, annular objective, and scanning optical microscopy, 107
Leukocytes
  differential counts, image analysis, 91–92
  Feulgen absorption cytophotometry, 180–181
Light
  harmfulness to cells, 28, 42–43
  photobiology, fluorescent photosensitizers, 67
Lilium, Feulgen absorption cytophotometry, 147
Livermore Laboratories, CYDAC, 95
Locusta migratoria, Feulgen absorption cytophotometry, 156
LSD, cathodoluminescence SEM, 265, 266–268, 271, 273, 275, 281–285
  incorporation, hair, 266–269, 281–285
Lucifer yellow, microspectrofluorometry, 59
poly-L-lysine coating, scanning microscopy, simultaneous light and electron, 122
Lysosome formation, GERL cisternae, high-voltage EM, 245, 247–251

Magnesium, X-ray microanalysis, nuclear ionic environment, 234
Malate, microspectrofluorometry, 47, 50–51
Malignant cells, cf. normal, microspectrofluorometry, 57–58
MDH, 51, 53
Meal worm, ultrasoft x-ray imaging, 305, 308, 310

338 / Index

Meiotic maturation, *Xenopus*, X-ray microanalysis, nuclear ionic environment, 234–236
Membrane-bound organelles, quantitative VIM, 73–74; *see also* specific organelles
Meningioma, human, cathodoluminescence SEM, 277, 278
Metabolic control. *See* Microspectrofluorometry, cell metabolic control
Metaphase chromosomes, high-voltage EM, 252–253
Methylene blue and methyl green, microphotometry, DNA and chromatin structure, 190
Microdensitometry quantitation, ultrasoft x-ray historadiography, 293
Microinterferometry, computer-assisted, living cell, 27–43
   cell volume determination, 28, 39–42
   RIC, 40, 41, 43
   Xenopus epidermal cell, 40
   densitometry (IBAS system), 29, 30, 40
   dry mass determination, 28, 29
   flow diagram, 35–36
   mitotic spindle formation, 32
   optical density, 38
   Psammechinus miliaris zygote, 29–32
   tissue culture cells, nuclei and nucleoli, 32–40
   Xenopus epidermis cell, 40
   Xenopus heart endothelial cell, 39
   image analysis, 29, 34–35, 42
   Leitz Mach-Zehnder type interference microscope, 29, 33–34, 37
   light, harmfulness to cells, 28, 42–43
   optical path differences, 27, 32, 41, 42
   refractometry, 28, 32, 39, 41
Microphotometry, nuclear structure and composition, 187–202
   chromatin, 187–190
      condensed, 188–190, 193, 197
      DNase I or II sensitivity, 188
      models for study, 188–189
      mouse hepatocytes, 189
      mouse thymocytes, 189
   chromatin protein demonstration methods, 197–200, 202
      listed, 202
   coefficient of variation, 196
   demonstration of DNA and chromatin structure, fluorescent probes, 190–197, 200
      listed, 201
      pre-stain manipulation of DNA, 196
   image analysis, 188
   Leitz MPV2 system with CELANA software, 189
   photomultiplier tube, 188
   transcription, 188
   video, 188
Microspectrofluorometry, cell metabolic control, 45–68
   adenylate charge, L cell response to malate, NADPH, 51–57
   cAMP, 65
   flavins, 45, 46, 59
   fluorescent carcinogens, intracellular interactions, 66–67
   G6P, 47, 48, 59
   malate, 47, 50–51
   metabolic compartmentation, NADPH, 48–51
   mitochondria, 47, 51, 53, 57–58
   modulation of intercellular flow, 59–65
      cell-to-cell inhomogeneities, 59, 63
      malignant ascites cells, 65
      multicellular integrated states, 59, 65
      pancreatic islet cells, monolayer culture, 60–64
      tracers, 59
   normal cf. malignant cells, 57–58
      free and bound coenzymes, identification, 58–59
   PFK, 60, 64
   photobiology, fluorescent photosensitizers (furocoumarins), 67
   sequential injection approaches, 45–47
Microspectrofluorometry, image analysis, 91
Microspectrophotometry. *See* Feulgen absorption microspectrophotometry (cytophotometry)
Microtubules
   AVEC-DIC analysis, 13–14, 23
      linear elements (MTLEs), 16, 18–21, 23

quantitative VIM, 73
Millipore filter, 116–117
Mithramycin, microphotometry, DNA and chromatin structure, 190
Mitochondria
  high-voltage EM, 249–255
  microspectrofluorometry, 47, 51, 53, 57–58
  quantitative VIM, 73, 79, 82
Mitochondria, wet elongated, potato, segmentation into spheres, EM, 313–324, 328
  cf. critical-point drying technique, 313, 314, 316, 319–323
    bleb-like protrusions from outer membrane, 319, 322, 323
  cf. filamentous, 319
  cf. fungal polysaccharides, 316, 318
  glow discharge, 314, 315
  sensitivity of mitochondria to storage and handling, 313–314, 323
  wet replication technique, 314–316, 328
    schema, 315
Mitosis
  chromosomes, cf. interphase, X-ray microanalysis, nuclear ionic environment, 223–224, 236
  high-voltage EM, 250, 252–256
  spindles
    HVEM, 254, 255
    microinterferometry, computer-assisted, 32
    quantitative VIM, 73, 80–83
Molluscum contagiosum, ultrasoft x-ray historadiography, 290
Mottle
  AVEC-DIC analysis, 14
  ultrasoft x-ray imaging, 304
  video-enhanced microscopy, 7
Muscle
  Chironomus striated, myofibrils, ultrasoft x-ray historadiography, 292
  diaphragm, rat, endocytosis, high-voltage EM, 245–248

NMR, 231, 232, 234
Notophthalmus viridescens, Feulgen absorption cytophotometry, 141, 146, 149–150

Nuclear magnetic resonance, 231, 232, 234
Nucleolus
  high-voltage EM, 256
  Xenopus oocyte, 232, 233
Nucleus/nuclear
  envelope
    high-voltage EM, 244, 245, 250, 252–253
    quantitative VIM, 73, 80
  point absorbance, Feulgen absorption cytophotometry, 177–179
  volume, Feulgen absorption cytophotometry, 152
  see also Microphotometry, nuclear structure and composition; X-ray microanalysis of nuclear ionic environment, electron probe; specific components

Olivomycin, microphotometry, DNA and chromatin structure, 192
Oocytes. *See under Xenopus laevis*
Optical beam induced contrast (OBIC) method, 111–112
Optical density
  Feulgen absorption cytophotometry, 169, 179
  microinterferometry, computer-assisted, 38
Optical error, Feulgen absorption cytophotometry, 167, 169–174, 177, 179–181
Optical path, microinterferometry, computer-assisted, 27, 32, 41, 42
Ovarian tumor mutants, *Drosophila melanogaster*, polytene chromosomes, 139, 140, 143, 152
Ovary, sea urchin, ultrasoft x-ray historadiography, 294

Pancreatic islet cells, microfluorometry, 60–64
PAP stain, 97–98
Pediastrum boryanum, Feulgen absorption cytophotometry, 147
Peripheral blood mononuclear cells, scanning microscopy, simultaneous light and electron, 125–127
Peroxisomes, high-voltage EM, 249

PFK, microspectrofluorometry, 60, 64
Phospholipids, 323
Photodiode, scanning optical microscopy, 104
Photomultiplier tube
  cathodoluminescence SEM, 265, 269, 270, 285
  Feulgen absorption cytophotometry, 142, 147, 150
  microphotometry, nuclear structure and composition, 188
Photon flux, AMACIIIS electro-optical flow transducer, 214
Photonic microscope system, video-enhanced microscopy, 8–10
Plasma membrane, high-voltage EM, 245, 246, 250, 252–253
Plastids, high-voltage EM, 254, 255
Plethedon glutinosus, Feulgen absorption cytophotometry, 141, 146, 149
Poecilia formosa, Feulgen absorption cytophotometry, 140, 145
Pollen, simultaneous light and electron scanning microscopy, 130–131
Polyclonal antibodies, fluorescent probes, Ca^{++} role in cell division, 74, 75, 83
Polysaccharides, fungal, 316, 318
Polytene chromosomes, *Drosophila melanogaster*, Feulgen absorption cytophotometry, 139–140, 143, 152
Potassium, cathodoluminescence SEM, 277; see also X-ray microanalysis of nuclear ionic environment, electron probe
Preparation, 97–98
Proflavine microphotometry, DNA and chromatin structure, 193
Propidium iodide
  AMACIIIS electro-optical flow transducer, 215–216, 219–221
  microphotometry, DNA and chromatin structure, 195
Psammechinus miliaris (sea urchin) zygote, microinterferometry, computer-assisted, 29–32; see also Sea urchin
Pseudemus scripta, Feulgen absorption cytophotometry, 140
Pseudonurse cells, Feulgen absorption cytophotometry, 139–140, 143, 152

PUFF-type scans, ultrasoft x-ray imaging, 300
Pyronin Y, microphotometry, DNA and chromatin structure, 195

Quinacrine dihydrochloride, cathodoluminescence SEM, 265
Quinacrine mustard, microphotometry, DNA and chromatin structure, 193–194

Raji cells, AMAC IIIS electro-optical flow transducer, 216, 220–221
Rana
  esculenta, Feulgen absorption cytophotometry, 141
  pipiens
    keratocytes, AVEC-DIC analysis, 14, 19, 23
    X-ray microanalysis, nuclear ionic environment, 225
Reflection interference microscopy, cell volume determination, 40, 41, 43; see also Microinterferometry, computer-assisted, living cell
Reflections and conclusions, 327–329
Refractometry, microinterferometry, computer-assisted, 28, 32, 39–41
Reproducibility, quantitative VIM system, 77
Resolution
  ultrasoft x-ray imaging, 304
  scanning optical microscopy, confocal, 106–107, 109
Rhabdomyosarcoma CCL 136, microspectrofluorometry, 57–59
RNase pretreatment, microphotometry, DNA and chromatin structure, 190, 192–193
RNP matrix, 234
Root apex, high-voltage EM, 249, 252–253
Rubidium, X-ray microanalysis, nuclear ionic environment, 229

Salmo gairdineri iridens, Feulgen absorption cytophotometry, 140–141, 143–146, 148, 154, 155
Saltation, AVEC-DIC analysis, 13
Salt gland cells, seagull, 58
Scanning electron microscopy, cathodoluminescence. See Cathodoluminescence SEM studies

Scanning microscopy, simultaneous light and
electron, 115-131
applications, 125-131
autofluorescence, 130, 131
botany and geology, 129
cancer markers, 125, 126, 128
immunocytochemistry, 129
PBMN cells, DNA measurement,
125-127
pollen, 130-131
description of combined instrument,
117-121
light and electron paths, 120
schema, 119
specimen placement, 119-120
vacuum chamber, 118-120
LM, internal structure, 116
SEM, surface, 115, 126
cf. sequential LM and SEM, 116-117
specimen preparation, 121-124
centrifugation, bucket, 122-124
conductivity, 121, 126
cover-glasses, 121
Feulgen DNA stain, 124-127
poly-L-lysine coating, 122
sedimentation method, 122
Scanning optical microscopy, 103-113
detectors, split detector mode, 110
laser, 104
optical beam induced contrast (OBIC)
method, 111-112
cf. EM, 111-112
semiconductor imaging, 111
photodiode, 104
type 1 (conventional), 104-105, 107
harmonic microscope, 112
type 2 (confocal), 105, 106, 327-328
annular objective lens, 107
axial scanning, 108-109
depth discrimination, depth of field,
107-110
improvement in resolution, 106-107,
109
Scanning transmitted electron
microscopy, 226, 233
Sea gull salt gland cells, 58
Sea urchin
eggs, quantitative VIM, 74, 82-83

ovary, ultrasoft x-ray
historadiography, 294
zygote, 29-32
Sedimentation method, simultaneous light
and electron scanning microscopy,
122
Semiconductor imaging, 111
Serial reconstruction, three-dimensional,
243-244
Sodium. *See* X-ray microanalysis of nuclear
ionic environment, electron probe
Sperm, *Drosophila melanogaster*, Feulgen
absorption cytophotometry, 140, 145,
151
Spindles. *See under* Mitosis
Stain/staining, 97-98
darkness, insufficient, Feulgen absorption
cytophotometry, 176-177
intensity, Feulgen absorption
cytophotometry, 168-169,
174-176, 181, 183
see also specific stains and dyes
Stereology, 90
Synchotrons, ultrasoft x-ray
historadiography, 296-297

TCA, microphotometry, nuclear structure
and composition, 198, 200
Thermonuclear fusion, ultrasoft x-ray
imaging, 299
Thiazin dyes, pH 4.0, microphotometry, 190,
194
Thioflavine T, cathodoluminescence SEM,
265
^{3}H-thymidine incorporation, correlation with
Feulgen absorption cytophotometry,
182-183
Thymocytes, Feulgen absorption
cytophotometry, 170-173, 175-177
Tissue culture cells, computer-assisted
microinterferometry, 32-40
Toluidine blue, microphotometry, 190,
198-199
Tradescantia paludosa, Feulgen absorption
cytophotometry, 147
Transcription, microphotometry, nuclear
structure and composition, 188
Transducers, AVEC-DIC analysis, 23
Trichoderma viride, 316, 318

Tunneling hypothesis, blue shift, microspectrofluorometry, 66

Ultrastructure, 327, 328
 AVEC-DIC analysis, 22–23
 high-voltage electron microscopy, 328; *see also* High-voltage EM, ultrastructural histochemistry
 ultrasoft x-ray, 328
 x-ray microanalysis, 328

Vacuum chamber, scanning microscopy, simultaneous light and electron, 118–120
Vicia faba, Feulgen absorption cytophotometry, 147
Vickers photomultiplier tube, Feulgen absorption cytophotometry, 142, 147, 150, 169, 177, 184
Video
 intensified, image analysis, 89–90
 microphotometry, nuclear structure and composition, 188
 see also AVEC *entries*
Video-enhanced microscopy, cell motility, 3–10
 AVEC methods, 5, 8, 9, 10
 differential mode, 9–10
 filtering mode, 10
 frame memories, 7–8
 image processing, digital, 3
 microscope design changes, 5–7
  homogeneity of illumination, 7
  image quality, 6
  mechanical stability, 6
  mottle, 7
  Zeiss Axiomat, 6–8
 photonic microscope system, 8–10
 sequential subtraction to detect motion, 9
 trace mode, 9
 video-intensification microscopy, 4–5, 9, 10
Video intensification microscopy, quantitative, Ca^{++} role in cell division, 73–84
 actomyosin-containing contractile ring, 73, 80
 ATP, 82–83
 calmodulin, 73

fluorescent probes
 chlortetracycline, 74, 78–81, 83
 polyclonal antibodies vs. Ca pump enzymes, 74, 75, 83
Haemanthus, 81, 82
HeLa cells, 78–83
mammalian cells in culture, 74
membrane-bound organelles, 73–74
 ER, 73, 80
 mitochondria, 73, 79, 82
 nuclear envelope, 73, 80
mitotic spindles, 73, 80–83
MTs, 73
QVIM system, 75–77
 image analysis, 75–76
 reproducibility, 77
sea urchin eggs, 74, 82–83

Water content, *Xenopus* oocyte, X-ray microanalysis, 231–234
Wet replication technique, 314–316, 318
 schema, 315
White blood cells, human, Feulgen absorption cytophotometry, 180–181

Xenopus laevis, 202
 Feulgen absorption cytophotometry, 138, 139, 141–148, 154–158
 microinterferometry, computer-assisted epidermis cell, 40
  heart endothelial cell, 39
 oocytes, X-ray microanalysis, nuclear ionic environment, 225, 227
  changes during meiotic maturation, 234–236
  free vs. bound K^+, 228–230, 235–237
  water content, 231–234
 ultrasoft x-ray historadiography, endothelial cell nuclei, 296
X-ray historadiography, ultrasoft, history, 289–297
 Chironomus striated muscle myofibrils, 292
 distance squared effect magnification, 293, 295
 dry mass, 289, 295–296
  Xenopus laevis endothelial cell nuclei, 296
 eclipsed by interference microscopes, 295

general design schema, 291
human epidermis with *Molluscum contagiosum*, 290
microdensitometry quantitation, 293
pin-hole, 289, 290
sea urchin ovary, 294
synchotrons, 296–297
X-ray imaging, ultrasoft, 299–310, 328
applications, 305–310
bee in flight, 305, 306
beetle, 305, 307
digitized scans, 309, 310
house mite, 309
meal worm, 305, 308, 310
chemical analysis techniques, 300
copper target, 301–303
dense plasma focus, 299, 300
PUFF-type source, 300
densitometer tracing, 303
laser fusion, 299, 300
Nd laser, 300, 301, 303
resolution, diffraction and mottle, 304
and thermonuclear fusion research, 299
X-ray conversion efficiencies, 302
X-ray microanalysis, 328
X-ray microanalysis of nuclear ionic environment (Na^+, K^+), electron probe, 223–237
Ca^{2+}, 225, 234–236
changes during meiotic maturation, *Xenopus* oocytes, 234–236
amiloride, 235, 236
GVBD, 234–236
chromatin structure, 223–224
condensation, 224
Cl^-, 229, 234, 235
dry weight vs. chemical activity, 225
extent of free and bound K^+, *Xenopus* oocytes, 228–230, 236–237
H^+, 234
ice crystal size as indicator of water content, *Xenopus* oocyte, 231–234
NMR relaxation times, 231, 232, 234
nucleolus, 232–233
RNP matrix, 234
interphase cf. mitotic chromosomes, 223–224, 236
Mg^{2+}, 234
quantitative energy dispersive, 229
Rh^+, 229
STEM, 226, 233
validation cf. known, accepted wet chemical measurements, 225–228
cryomicrodissectioin, 226, 227
Hall mass fraction method, 226

Zea mays, Feulgen absorption cytophotometry, 147
Zeiss
Axiomat
AVEC-DIC analysis, cytoplasmic transport, 14, 23, 24
video-enhanced microscopy, 6–8
microfluorospectrometry, image analysis, 91
Zygote, *Psammechinus miliaris*, microinterferometry, computer-assisted, 29–32

DATE DUE